Challenging the Safety Quo

Safety is broken. The people who are responsible for helping you stay safe should be at the top of your Christmas card list, but all too often they are despised, ridiculed and ignored.

But safety management is beginning to be challenged. Businesses have begun to realise that what they have been doing is no longer providing any additional value.

The same issues are repeatedly raised by corporate leadership:

- How do we get our workforce engaged in safety?
- How do we improve safety systems to gain commitment from all employees?
- How do we improve safety understanding to make the case for change?
- How do we embed safety as an integral part of culture in an environment of ongoing change and cost pressure?

Challenging the Safety Quo makes the case for change based on stagnating performance, identifies areas where there are problems and proposes alternative ways to progress. Provocative but practical, it outlines the business benefits to be gained from putting in place the right approaches to managing safety, although not in the way traditionally presented by most safety managers.

This book translates theory into practice; putting an accessible, practical and usable spin on cutting-edge thinking in safety.

Craig Marriott is a senior safety professional with over 25 years' experience managing safety in high-hazard industries. From nuclear submarines and highly radioactive waste, to high-pressure gas pipelines and oil rigs, he has written safety cases and managed safety for some of the world's most hazardous operations. Craig specialises in providing safety leadership advice to organisations across the world and has worked with companies in the US, Canada, Europe, Middle East and Asia Pacific.

Craig provides a refreshing perspective, moving past the bluster and pointing toward a more productive future for safety management. His approach challenges old assumptions, but in a practical way that gives us a way forward.

Ron Gantt, *Vice President, SCM*

Challenging the Safety Quo

Craig Marriott

Routledge
Taylor & Francis Group

LONDON AND NEW YORK

First published 2018
by Routledge
2 Park Square, Milton Park, Abingdon, Oxon OX14 4RN

and by Routledge
711 Third Avenue, New York, NY 10017

Routledge is an imprint of the Taylor & Francis Group, an informa business

British Library Cataloguing in Publication Data
A catalogue record for this book is available from the British Library

Library of Congress Cataloging in Publication Data
Names: Marriott, Craig, author.
Title: Challenging the safety quo / Craig Marriott.
Description: New York : Routledge, 2018. | Includes bibliographical
references and index.
Identifiers: LCCN 2017036688| ISBN 9781138558748 (hardback) |
ISBN 9781138558762 (pbk.) | ISBN 9781315151397 (ebook)
Subjects: LCSH: Industrial safety–Management.
Classification: LCC T55 .M3555 2018 | DDC 658.4/08–dc23
LC record available at https://lccn.loc.gov/2017036688

ISBN: 978-1-138-55874-8 (hbk)
ISBN: 978-1-138-55876-2 (pbk)
ISBN: 978-1-315-15139-7 (ebk)

Typeset in Bembo
by Wearset Ltd, Boldon, Tyne and Wear

Contents

Figures

Tables

Part I
Introduction

1 The problem

On 19 November 2010, an explosion ripped through the Pike River Mine in the South Island of New Zealand. A second followed soon after.

Twenty-nine men died.

Fathers. Brothers. Sons.

In any such disaster, more lives are ruined than those that are taken directly. In a small community such as this, the impact was particularly devastating (Macfie, 2013). The disaster triggered a response that had a much broader impact than simply on this mine or this industry.

The safety record of New Zealand's industry was already under scrutiny due to perceived poor performance in comparison with other developed countries. In the wake of Pike River, industry scrutiny gave way to a public clamour for change, for improvement and for accountability. The government duly obliged.

An independent task force was established to review industry performance across all sectors. A Royal Commission investigated the Pike River tragedy. The regulator was shaken up, given more independence and more resources. Legislation was changed. Targets were set and promises made. It took five and a half years for new legislation to be enacted and come into force.

Will it work?

Who knows?

Perhaps (and hopefully) it will, but history suggests otherwise. Similar, previous tragedies triggered similar responses and similar issues are playing out all around the world as various governments and agencies attempt to come to grips with the problem. In the UK, Flixborough prompted changes in the 1970s; Piper Alpha prompted changes in the 1980s. Union Carbide's facility in Bhopal provoked a similar outcry. And there are many others – Texas City, Longford, Deepwater Horizon, Chernobyl, Seveso...

So this has happened over and again, for the last 40 years. Yet here we are, still failing to provide safe workplaces for our colleagues. Nor is this confined to major disasters with multiple fatalities. While general accident rates across industry have been in general decline for some time, serious injuries and fatalities have not shown a similar reduction – either reducing much more slowly, plateauing or, in some cases, increasing.

Why is that? Based on the rhetoric after each tragedy, these issues should be sorted out by now:

> 'We must do better; bad practices will be weeded out and poor performers prosecuted under the new laws…'

> 'We will look further and delve deeper to root out the bad apples…'

> 'We will hold up, reward and recognise good performers; identify best practices and share them.'

But, do we actually know what best practices look like? What if our current view of what is good is completely wrong? What if we're looking in all the wrong places?

These questions are currently being asked and the safety status quo – the *safety quo* – is being challenged. Unseen by most of the outside world, there is a debate raging in the safety profession. In online forums, at conferences, seminars and in safety publications, people are beginning to question some of the most well-established principles of safety management. Safety as a profession is going through a mid-life crisis.

Some of the theories behind these challenges are well developed, particularly in academia, but in industry many organisations are very much behind the times and haven't yet got up to speed. Other theories are cutting edge and leading to howls of indignation from defenders of the safety quo, but are gradually gaining support as a few lone voices become a groundswell of opinion.

This change is needed because the truth is that safety is broken. There are pockets of excellence, there are good-quality people working very hard and with the best of intentions. We have made significant progress since the days when major projects budgeted for a certain number of fatalities. But, by and large, the safety profession is frowned upon. What is the general response across industry (or in your business) when the safety person gets involved in a conversation, or arrives at a site to observe work? Or when a new revision of the safety paperwork is issued? Are these welcomed with open arms? Are they seen as a positive inclusion to keep us all in one piece? No, they are not. There is a general rolling of eyes and resigned shrugging of shoulders, if not outright hostility.

How did we come to this? Everybody at work wants to go home fully intact. Nobody wants to die on the job, or be seriously injured. How did we get to the stage where the people charged with helping to support that most fundamental of requirements – staying alive – are almost universally disparaged? On face value, it seems remarkable. Businesses ask how they can make their workers more engaged in safety. But it is surely the natural state for people to be engaged in their own safety. In fact, it is an evolutionary imperative. The question business should be asking is, therefore, *why* are they disengaged and what is the business doing to contribute to it? Once that is answered, they can stop those activities that are actively disengaging their workers from safety.

Of course, safety is not the only department that strikes fear into the heart of the rest of the business. Procurement, accounting, IT and legal also tend to have the same chilling effect. But this is largely a function of the bureaucracy of big business and differs in two ways. First, they tend to impact more on the back-office staff who are at least paperwork-savvy rather than the front-line with their almost pathological distaste for reams of paper. Second, disengagement in the process of dealing with these departments usually has little more effect than some delays and productivity inertia. Damaging to the business perhaps, but not disabling to individuals.

To re-balance and to improve our safety performance we need strong leadership – this is something that everyone agrees with. And to paraphrase Peter Drucker, leadership is not about doing things right, but about doing the right things.

About this book

This is not an academic text. It does not consist of exhaustively researched reference information (although there are some references to other reading of interest). It does not even purport to be right. I have included some aspects that I am not entirely convinced by personally, but which are views held by others whom I respect. Its principal intention is to encourage critical thinking about safety by challenging some of the fundamental assumptions we are currently using that don't appear to be working as well as we need. It also offers some alternatives that may be of use.

It is, however, based on my experience of over 25 years in the safety field, the majority in the nuclear and petrochemical industries – high-hazard industries where consequences of failure are severe and where safety vigilance and performance ought to be (and in most cases are) industry leading.

The book covers a wide range of topics as an introduction to some of the thinking that has the potential to improve our safety performance. It will serve as a gateway to other resources, researchers and thinkers who have presented some of that thinking in more depth in their own work. There is a gap between those at the forefront of safety thinking and those at the fore front of safety doing. Hopefully, we can start a discussion that begins to close that gap. So, think of it less as a training resource and more as a chat over coffee about the state of safety management.

Note that we talk about safety throughout this book. The challenges and lessons apply equally (if not more) to occupational health. Wherever we use the word 'safety' it should be read as 'health and safety'. I am not well enough versed in the intricacies of occupational health to provide a detailed critique, but health and safety are so often put together that I have a working knowledge, although I am as guilty as anyone of forgetting the health in health and safety. The single biggest change we can make to improve occupational health (apart from increasing awareness) is to start treating it as a genuine risk and attempt to manage it at source, rather than our current approach of (at best) monitoring its effects. I encourage anyone who is a specialist in this area to consider the concepts included here and challenge our current approach to occupational health as well.

The book is in three parts:

Part I Introduction.
This is self-explanatory. It introduces the problem and some of the background and context to it.

Part II Truths, half-truths and downright myths.
This is where we challenge assumptions and generally poke holes in things. It includes a selection of aspects that highlight some of our current misapprehensions.

   It has become somewhat fashionable to publish business books that don't need to be read in the order in which they are written. You can jump about and dip in and out. In this case, the chapters within Part II have no particular order to them. They are a selection of approaches used within safety that are worthy of being challenged. But they were written in order, so there may be some back-referencing that you will miss if you choose to be a bit ad hoc. But if there is a particular area that interests you, feel free to go straight for it. I'm sure you'll work it out.

Part III Your context
Good safety is hugely context-specific. Part III is as close as I get to telling you how to do it. But really, it's about how to understand your business and where safety fits into it. This is to give you a fighting chance of making sustainable improvements.

So, I'm not here to give you all the answers, but to give you good questions to ask. If you were hoping for a quick fix, then sorry, but that's the first myth to dispel. The silver bullet doesn't exist, whatever the claims of that latest system or programme that someone is trying to sell you.

So, read on and respond in any way you see fit, as long as you can logically and objectively support your position. Argue against it, agree with it, vehemently oppose it, denounce or support it…

   … but don't ignore it. There are lives at stake.

Note

You can provide feedback or comments at www.safetyquo.com.

Reference

Macfie, Rebecca (2013) *Tragedy at Pike River Mine*, Awa Press.

2 The cast list

There are three fundamentally important groups of people in the delivery of safe operation:

1 management
2 workers
3 safety professionals.

For the purposes of this book, management encompasses all those people that have some formal degree of influence over the working environment – from the board, through the executive suite, middle management and to supervisory level. When I refer to 'management' or a 'manager' in a generic sense it can be any of these. There is a difference between managers and leaders, which we will explore in a later chapter, that is the subject of an entire industry of performance improvement through better leadership. But the 'management' in this context undertakes both leadership and management activities and is more of an organisational hierarchy definition than a behavioural one.

All employees are workers (well, most of the time), but we are taking them to be those who are the most exposed to the hazards of the workplace, i.e. 'front-line' workers, be they miners, joiners, forklift drivers, tree fellers and so on. This means there can be some overlap with certain individuals in both the worker and management category (for example, a scaffold leading hand), but this should make no material difference to the discussion.

Safety professional means just that – those people who earn their living by their safety work, whether as an employee, consultant or academic. Although there are some similarities at a basic level, it is not intended to include a workforce safety representative in this description.

Different organisations have differing levels of safety maturity – as may different departments and teams within a single organisation. The relative importance of each of these roles varies with that maturity and there is a high degree of inter-dependence. The influence of a safety professional, for example, can be minimal if there is little support, or even downright opposition, by management to their attempts. More on safety maturity later.

Management

The role of management in delivering safe operation has been increasingly recognised over the last few years. There are a number of activities that management can undertake that facilitate this. Conversely, not doing these, or getting them wrong, can hinder or prevent safety altogether. Such activities are routinely listed in safety books and training courses and are, when considered at this level, broadly correct. They include:

- setting a clear and unambiguous safety-driven strategy;
- demonstrating commitment to safety;
- supporting a strong safety culture;
- being personally involved in safety;
- getting out in the field and listening to the workforce;
- active involvement in investigations and recommendations;
- encouraging incident reporting;
- rewarding good safety performance;
- providing adequate resources to support safety.

In short, say that safety is important and then show it – always.

If you are reading this and are not a manager, please note that despite what some people (and some successful cartoonists) may suggest, there is not a soul-removal process in the recruitment of managers into industry. Management, on the whole, is not evil. This may come as a surprise, but experience shows it to be the case. I have not yet come across the manager who would accept, let alone welcome, a serious injury or fatality purely in the pursuit of higher profits. Managers can be distant; preoccupied with other matters; unaware of their impact on safety; lacking the knowledge to make it work; uninformed of the risks being undertaken and subject to a whole host of other inadequacies, but they do care. One only has to watch a few episodes of *Undercover Boss* to see how horrified senior managers are when they find out what the real world is all about. When asked, all managers I have ever worked with will answer that they are committed to the safety of their staff. Of course, they are not going to say anything else in public, but they genuinely mean it. They just don't always do well at sharing that commitment.

However, if you are reading this and you *are* a manager, when you say you're committed to safety, do you know what that means? Are you actually committed? Do you demonstrate that consistently? Stop for a few minutes and answer the following questions:

> I think I'm committed to safety. But what does that actually mean in practice? What clearly visible (because that's important) activities do I routinely and consistently do that demonstrate that commitment and contribute to the safety of my workforce? Do I fully understand how these actions translate into safety benefits? How does what I say, ask and

do impact on others within my organisation and what are the safety consequences of that impact?

The fact that you have chosen to read this book suggests you will be able to provide some answers. Hopefully, by the end you will be in a position to pinpoint which areas of your safety thinking you would like to develop some more. If you're struggling, there are some clues in Appendix 1 and we'll talk more about management and leadership later.

Workers

The workers bear the brunt of safety problems. After all, it is they who are most exposed to the hazards present. There are many things that can (and should) be done in planning, preparation, design and systems to prevent the workers being unduly exposed, and these are hugely important (see *managers* and *safety professionals* sections above and below). But it is worth noting that a common theme in most major industrial accidents is one of an unforeseen or unexpected combination of events leading to the catastrophe. If we can't predict or foresee them, we can't build appropriate responses into procedures and rules. Given this, it is of the utmost importance that the workers are risk-aware, understand the systems and equipment that they are using and have both the capability and autonomy to respond to unusual, often rapidly evolving, situations in the right way (I use 'autonomy' and avoid the word 'empowerment' as this is too much of a buzz-word with blame-passing connotations).

Workplaces are ever more complex environments and what we see in our design drawings and process manuals can misrepresent vital real-life situations and interactions. Particularly where equipment is old or unusual, or where there are multiple interfaces and overlaps, hands-on operational experience is crucial to understanding the risks. The role of the workers in providing this information must never be underestimated or ignored. Given the complexity and often competing pressures they face, the workers find ways to carry out the job effectively, in spite of the approved way of doing it, rather than because of what is laid down in procedures. Hollnagel (2014) refers to this as 'work as done' as distinct from 'work as imagined'.

This must be balanced, though, with an understanding that the workers' knowledge is limited to their own career experience. One of the most common arguments against a new approach to an activity is, 'I've been doing this for thirty years and never hurt myself yet.' But absence of injury does not equal presence of safety (this is a crucial consideration we will return to later). Fatal accidents do not occur once per career – we would have very short careers, if so. Typical statistics show, for most countries, fatalities of the order of one per 10,000 to 100,000 workers per year. Most of those people will also have had a healthy history of successfully working without hurting themselves. As they say in financial services, past results are no guarantee of future performance. It is entirely possible that their operational understanding of the risk is based on undertaking an activity unsafely, but unhurt, over a period of time, providing them with an anecdotal impression of adequate risk management. Combining their intimate knowledge with a fresh, outside perspective is a difficult but crucial balance to achieve. Most organisations err heavily on the side of the outside perspective and fail to account for worker knowledge.

Far too often, workers are blamed for accidents. The vast majority of accident investigations still conclude with a statement of 'The worker did ...' or, 'The worker failed to ...'. Even in the aviation industry – generally at the forefront of good safety thinking – the number of crash investigations that conclude 'pilot error' is astonishing. The worker should not be thought of as the individual solely responsible for safe outcomes. They have a part to play, but should be thought of as the sharp end of a large organisational tool we use to complete work. That tool is contributed to and deployed by the entire organisation – from management through design, procurement, accountancy, human resources, maintenance and so on. Accidents occur due to failings in this whole safety supply chain.

If you are reading this and are not a worker, please note that there is actually no requirement for workers to leave their brains at the factory gate. They do not need to be treated like children with no understanding of what is going on around them and force-fed simplistic rules. Listen to them and respond to their concerns. They know many of the risks better than you do. That's not to say they're always right, of course, but when they're not, an explanation as to why is needed. Yes, there may be occasional individuals who will seem to be hard-wired to be obstinate, argumentative and anti-management, but these are actually thin on the ground and the better the culture and environment, the fewer there will be and the less influence they will have.

If you're reading this and you *are* a worker, here are some questions for you:

> Do I genuinely concern myself with safety, or do I honestly have an 'it won't happen to me' attitude. I know I have the right to stop work if it is unsafe, but would I have the courage to do so? Am I risk aware and understand the significant hazards that can cause me real harm? Do I think hard about my safety documentation, or do I 'tick and flick' to get it done?

There is a thread that will be returned to throughout this book. Among the myriad of complexities jostling for position in the safety importance hierarchy, there is one thing that stands out above all else – if the interest, buy-in and awareness of the workforce are lost, your safety vision will fail. If you have good quality in your other systems, it may take longer and fail less spectacularly than it might otherwise, but without the workers it *will* fail. There is no silver bullet, but a top-quality, engaged workforce is the closest thing to it that there is.

Safety professionals

Somewhere between the management and the workforce, sometimes trusted and supported, sometimes not, lives the safety professional. It can be a hard life, a conduit of antagonism from the shop floor to the upper echelons and back again.

Safety professionals come from a huge variety of backgrounds. Some are engineers and scientists, academics and psychologists developing ergonomic, behavioural and cultural models to understand the complex interplay in the workplace. Others are time-served workers who developed a passion for safety after years of exposure to the hazards. They know the facility inside and out, knowing who to speak to and which buttons to push to get buy-in and make progress. Unfortunately, a significant proportion are also carrying out the role as a part-time addition to other duties, or because they have outlived their usefulness elsewhere. In both of these cases, there is clearly little drive for good safety within the organisation, but this is exacerbated by the appointment of someone who is not a good fit for the role.

All too often, the safety professional is left to 'manage safety' and can be seen as a handbrake to progressing the task at hand. Their expectations are seen as extra cost, time and effort that is not value-adding.

But safety is an insurance policy. Like any other insurance, there is a price to pay up-front. It is a relatively small price, but it gets noticed. If everything goes well, it's easy to look at it and say that it was not worth paying. After all, who looks back over 20 years of paying insurance premiums for their house with satisfaction if they have never made a claim? But when the house burns down, how pleased are they that they did? With hindsight, insurance is always worth having where there is the potential for a large loss. Safety is the same (although there are those who claim that safety is not actually a cost, but an investment that will reap benefits – we will discuss this later on). The role of the safety professional boils down to helping the organisation to turn that hindsight into foresight, managing and carrying out the work so that the high price does not have to be paid.

If you are reading this and are not a safety professional, please note that it is not actually their role in life to make things stop, ban activities and generally act as a huge barrier to progress. The problem is, however, that many of them see no other way to go about their business. Safety is a complex business. It involves many people, many drivers, many interactions and many opposing opinions and requires the ability to manage all of them. Unfortunately, this complexity is little understood and this leads to a vicious circle in industry where safety as a profession is not valued, and therefore not always adequately remunerated to attract high-quality people (Figure 2.1). Performance is subsequently poor, which reinforces the idea that it's not difficult and anybody can do it. After all, how hard is it to make sure everyone does a risk assessment and look for people without hard hats on?

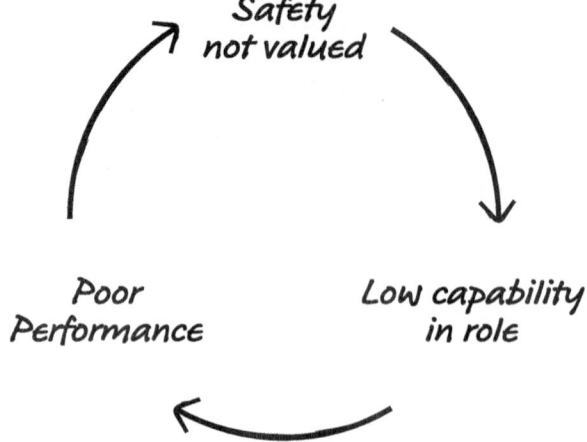

Figure 2.1 Vicious circle of poor safety support.

But the world, and industry, has moved on. No longer should the safety profession be dominated by ex-shop floor workers with a good knowledge of the process and what the immediate hazards are. The people with those skills are still required and their contribution is hugely valued, but at the cutting edge this is no longer enough on its own. You can still tinker with your car at home and change the brake pads when they are worn and it will make it significantly safer, but gone are the days when an enthusiastic amateur could do all vehicle maintenance. Now there are engine management systems – sophisticated computers that need purpose-built diagnostic tools. The electronic stability control has added an extra level of safety, but it needs an equal level of sophistication and capability to ensure it works and to improve it still further, even though there is still a need for the mechanic to tighten nuts and bolts. Likewise with safety. To move successfully to the next level, it requires a step change in thinking and in capability.

Good, modern safety professionals should, therefore, be better rewarded and better appreciated. Without this, the individuals with the necessary critical thinking capability will not be attracted to the profession. A top-quality safety professional is conversant with organisational culture; organisational psychology; individual psychology; motivational theories; basic medical requirements; an understanding of chemicals and their effects; probability and risk theory; ergonomics; engineering techniques across all disciplines; operational activities across a range of industries; the process impact of temperature and pressure; legislation and regulation; communication techniques and influencing skills from workface to board and everywhere in between; statistical analysis and a whole host of other complexities. On top of that, they have to be able to translate between all the different specialists in each of those areas. It is not a simple job.

If you are reading this and you *are* a safety professional, no doubt you are nodding sagely and agreeing with all of this, working out how you can build it into your next pay review. But, ask yourself these questions:

> Am I conversant with all areas that I should be? Do I engage at all levels? Do I research different approaches and understand their merits? Do I spend enough time in the field as well as the office? Do I carefully consider my workplace and develop safety systems that align with the hazards and risks present, or just regurgitate the same old requirements everyone does? Do I understand my role in the organisation and how my approach impacts on our performance? Can I look myself in the eye and say I am doing everything possible to keep my colleagues safe – sometimes in spite of themselves?

I look at some online safety forums and see some of the discussions on there that leave me worried for the future of the profession. Others agree – one particularly basic request from a so-called safety specialist led to the response: *'If you have to ask that, maybe you need to get out of the way and bring in a proper safety professional to do it.'* But I also see some of the sheer brilliance that is out there. There is some fascinating research material and thinking that will drive massive improvements in safety performance. Provided that it is allowed to spread and oust the entrenched thinking that is currently in place.

There is no lack of passion and desire across the majority of safety professionals and it is better to begin with that and harness it than it is to take the most perfectly developed safety solutions and try to get them applied and embedded by a disinterested part-timer.

If you are a safety professional, there may be a temptation to take offence at some of the discussions coming up where long-practised approaches are challenged. I ask you to be open-minded and accept that challenge. Many of those older concepts have served a purpose and achieved improvements – it is just that they may no longer be the right approach as we move forward into the complex and rapidly changing future of work. We don't all have to agree – no progress was ever made by everybody thinking the same thing. Any opinion, concept, idea or approach stands and falls by its robustness to challenge. A good concept rebuffs challenge successfully by strong counter-arguments and strengthens its own position. Challenge is necessary for progress and underpins the learning culture that is a hallmark of successful safety management.

Reference

Hollnagel, Erik (2014) *Safety-I and Safety-II: The Past and Future of Safety Management*, Ashgate.

3 The maturity relevance

The principal premise of this book is that we are clinging to old methods of safety management that are no longer working, although there are some more recent innovations that also appear to be misguided. The reason that these methods are in place, by and large, is because they have worked in the past. There is little doubt that safety performance is now better than it was 40, 50 or 100 years ago. Watch any of the videos that are widely available of workers building major projects in history such as the Empire State Building. It is amazing to see the labourers sitting on a steel girder hundreds of metres above the ground eating their lunch – no harnesses, no hard hats (although possibly a flat cap); nothing between them and a plunge to their deaths. Indeed, such projects used to budget for a number of fatalities in their planning, an approach that would be inconceivable today. The London Olympics in 2012 were the first Games to be delivered without a fatality in the construction stages, demonstrating how far we have come.

So, the challenges here are not intended to be too deeply critical of what has gone before. The point is that our safety maturity has reached levels where the majority of obvious fixes have been made. As with any endeavour, improving the final few per cent is the hardest. Engineering standards have been developed that deal with many risks at source. Robust work control mechanisms are in place. Audits and reviews help to provide continuous improvement. This is why the role of the safety professional has evolved into a much more complex and challenging one. In a perfect demonstration of the 'survival of the fittest' principle, only the toughest and most difficult to solve safety challenges remain.

At least, that is what we would like to think, and it is true for those organisations that are at the forefront of safety management. However, some still lag behind in their safety maturity.

As the name implies, safety maturity grows over time by building on experience and knowledge gained to gradually improve. Wisdom comes with age.

Organisational psychologists have developed a number of maturity models to describe this phenomenon. One such produced by the Keil Centre in Edinburgh, for example, identifies a number of steps in the maturing process.

Their *Safety Culture Maturity Model®* shows both an increasing skill and knowledge process, but also increasing inter-dependence among the organisational components, highlighting the importance of culture in more mature safety systems. Other models are also in place, but they all tend to have the same fundamental premise of increasing maturity moving up levels of a ladder, as shown in Figure 3.1. In such models the maturity of an organisation can be measured by aligning observable traits with the level descriptions. These are widely used and will be familiar to many organisations.

This is a fascinating subject in its own right and far more detailed than the routinely showed maturity ladders would suggest, which are typically presented without all of the underlying detail. Like many models, these are an over-simplification of reality and the idea of applying them too rigidly should be resisted.

Some more information can be found through the various published models, or in a UK Health and Safety Executive research paper (UKHSE, 2000). As a general aside, the UKHSE website (www.hse.gov.uk) provides some of the best and most comprehensive guidance available anywhere.

Safety culture development is a collaborative process and involves whole organisations and whole industries, with many interconnected factors involved. Different parts of the same organisation can have different levels of maturity, as can different individuals, but a general picture can be gained. The offshore oil industry, for example, has a more mature safety culture than does the residential construction sector. This is not true for all people and all companies in those sectors, but holds well at an overview level.

As noted above, some of the approaches that I argue against in this book have been successful in the past. Particularly for those organisations that may be on a lower rung of the maturity ladder, *some of them may still be effective*. I am not suggesting we throw them away for the sake of it. Every situation demands a different solution. This book is aimed primarily at those companies and people with a relatively mature culture, looking to continue to improve. In mature companies, for example, we are well past the stage where a number

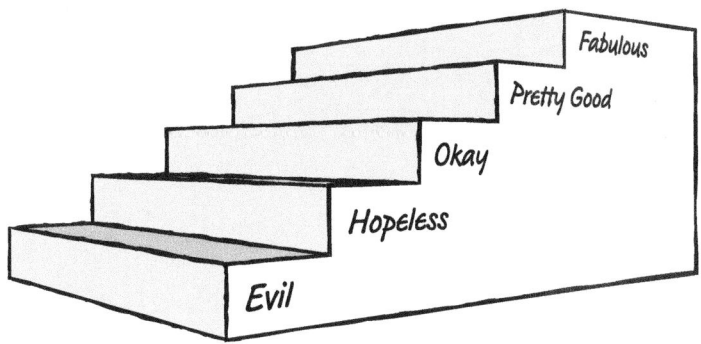

Figure 3.1 Safety maturity ladder.

of fatalities is a useful measure, or one we would even contemplate. Fatalities are so rare that a measure would almost always be zero, giving us no useful information, and any such event would cause shock waves that would reverberate around the business.

Unfortunately, there are some industries where the body count is high enough and variable enough to be able to discern improvement trends. In these industries, the old-school approaches may well still apply. I would hope that anyone in such a position would still find useful direction in this book. There is nothing in here that will derail efforts at this level and the approaches discussed should complement those more traditional ones if implemented correctly. Close attention should still be paid to the messages, language and approaches used now to make sure that nothing is being instilled as a habit that will be difficult to change when the more sophisticated thinking is applied later on.

Safety culture has become a hot topic of discussion, but unfortunately most of the discussion is relatively uninformed and overly simplistic. As noted above, culture is a highly complex area. When we get to Part III, we will discuss culture more in order to understand how to take safety performance forward. However, the key thing to bear in mind is that there is not really any such thing as safety culture (research papers notwithstanding) – there is simply culture. You cannot dissociate safety culture from the broader business culture. One is a subset of the other. They are not distinct concepts.

A good safety culture requires, among other things, openness, transparency and willingness to listen on the part of the management. It is simply not possible to have such a culture for safety in isolation, if it doesn't exist in the business as a whole. If business improvement ideas are quashed by a 'we know best' management team, people will not be forthcoming about safety ideas, no matter how many times they are told that it is a good thing.

If you want to improve your safety culture, it must be done in the context of improving (or at least understanding) the business culture. Safety is a good place to start your overall culture change programme, given its people-centric nature, but the two must go hand in hand. In Part III, we will use the business context to help drive better safety.

Reference

UKHSE (2000), *Safety Culture Maturity Model*, Offshore Technology Report, 2000/049.

4   The paradigm shift

We all talk about safety a lot. Typically we have a safety manager at work and (hopefully) a safety team to support them. We have safety committees and safety audits. We have safety regulators to make sure we're complying with legal requirements. We have safety days and safety campaigns. In the wider world there are safety standards for items that we purchase; social services to ensure the safety of children in the home; patient safety in hospitals; drug safety in pharmaceuticals and food safety when manufacturing and preparing edible commodities.

Safety is one of those words that is so commonly used and so familiar to us, that we rarely stop and think about what it actually means and whether we all attach the same definition to it when we use it. We assume we do, but is that the case? If we are going to challenge something, it is a fairly fundamental requirement to understand what it is.

Safety is typically viewed in quite transactional and absolute terms. If asked what safety in the workplace is, most people respond not with a definition of safety, but rather a litany of the things that we do in managing safety. It is paperwork; it is following the rules; it is doing what you're supposed to do. It is a *thing*, something concrete that you can picture while it is in progress. Usually, something that is done in addition to the actual task at hand – whether that be putting on safety glasses, writing a job hazard analysis or nominating a member of the team to act as a safety watch during the job.

When viewed in a context such as this, safety seems very straightforward. It is a pre-packaged box of activities that get applied to a job to stop people from getting hurt. In fact, many safety managers talk about using the right tool from their safety toolbox to approach a particular challenge. This defines safety as a separate activity from the job itself, something self-contained and simple that can be pulled out when required. By thinking in these terms, we restrict the potential for improvement. There is not a lot of point in challenging the role of a hammer in banging in a nail while building an item of furniture. We can make incremental improvements – a larger hammer head, a more ergonomic grip, a nail gun – or with slightly more lateral thinking could swap out the hammer and nail for a screw and screwdriver to get extra structural stability in the finished product. But we cannot make radical and fundamental changes to optimise the overall process while thinking at a transactional level. Genuine innovation and change comes from thinking at a higher level, considering the overall context and intent – by thinking conceptually about the whole.

To innovate, it is necessary to move away from the tools and the details and think about concepts and challenge ideas. Staying with our furniture metaphor, if we restrict ourselves to using our toolbox to make an improved cabinet for storing compact discs, there are a number of approaches we can take. We could make it bigger to fit more in; we could make it of different materials or with improved joint types to make it more robust or we could change the colours or the overall shape to make it more stylish. But if we think more broadly about conceptual aspects we get much greater scope to innovate and to change. Why do we have the storage unit? The purpose of our storage is to provide as much musical choice as possible to the listener. If we stop thinking about compact disc storage and start thinking about music delivery systems, the question becomes not how can we improve the compact disc storage, but how can we best deliver maximum choice in music? This opens up possibilities of storing music on hard drives instead of discs; of cloud-based storage of mp3 files; of wireless connection from a device in a different location; and so on.

It is important to note that thinking in this way does not necessarily preclude simpler thinking, or prevent us from reverting to our original plan for the storage unit. But it does provide us with many more options and opens up new horizons for better achieving our ultimate aim.

The above example is relatively simple and easy to think about given the real and solid components that are being considered – it is comparatively straightforward to think about what constitutes an improvement. Nevertheless, what constitutes an improvement is still subjective. Some people prefer the tactile nature of compact discs (or even vinyl records) to the somewhat ephemeral nature of cloud-based music storage. Where the concept under consideration is more abstract it becomes even trickier to determine what are improvements because it is harder to define an abstract concept.

Take the concept of *value* as an example. Again, this is something we use a lot and see a lot – great-value deals are advertised regularly, value meals are available at fast-food outlets and we strive for better value in negotiations in everything from home purchases to employment contracts. But what is value? At their first attempt, most people come up with a definition along financial lines related to getting more for less. But there are many different definitions of value. We sometimes use value as a euphemism for cheap; 'value' ranges at supermarkets are typically low-quality, low-price options – getting less for less. Sometimes we think of value as purchasing something at a discount – getting the same for less. We can also define good value as paying more for something which is of good quality and so will last for longer – getting more for more. Still others relate the value of something to their personal values – it is less about the price and more about the attribute that comes with it, whether it be quality, substance, content or brand.

Which all appears very confusing, yet we would all see these as normal and acceptable definitions of value. Therefore, it can be seen quite clearly that an abstract concept does not necessarily have to have a single, fixed and agreed-upon definition for it to remain useful. The definition can be fluid, taking into account the person involved, the context, the intended outcome and a range of other factors.

Safety can be considered as a similarly abstract concept. It, too, can have many nuances in definition. What may be considered safe for a front-line soldier would not be considered safe for a librarian; a flight of stairs that is safe for a parent may not be safe for their young child; and what was thought to be safe 50 years ago may well not be considered safe today. We shall expand on some of these ideas later on, but as we challenge some of the established ideas in safety, we need to begin to think of safety as a concept, rather than as the tools that we use and the rules that we follow. In doing this activity, what are we trying to achieve? What will be the outcome in different contexts? How will it interact with other areas? What is the underlying activity that we are undertaking and how does what we are doing help or hinder that activity? What unexpected or unintended consequences could arise from its application? Is this the only way to achieve the intended result? If not, is it the best way? How do we know? From whose perspective are we viewing this when we do it?

There is a need to make a paradigm shift from safety as a thing we do, to safety as a multifaceted concept. From something fixed and pre-determined to something that is flexible, subject to interpretation and context-specific. From something known to something worth exploring.

Even if you, personally, have a very rigid idea of what can be considered safe, thinking about it in this way can increase the routes available to you in trying to reach that particular safe destination. As we progress through this book, keep safety in mind as a concept that can be expanded, reduced, moulded, grown, dissected and rebuilt in line with a broader vision.

PART I SUMMARY

- History shows us that we don't appear to learn lessons from major accidents as well as we should.
- Safety performance has plateaued and the approaches that got us here may not be the right ones to continue to improve.
- Safety practitioners are not valued as much as they should be – but some of that is of their own making.
- Safety cannot be separated out from the broader business if it is to be successfully achieved.
- To innovate and improve, we need to think of safety as an idea open to interpretation, rather than a tool to apply.

In Part II, we will look at specific approaches routinely used in safety management and consider whether they meet the ideas and concepts outlined above.

Part II

Truths, half–truths and downright myths

5 The triangular fallacy

We might as well start with a good one.

If you have ever attended a safety course, or a company safety induction, it is almost certain that at some point you will have been introduced to the accident triangle. This is probably the most widely used descriptor in the safety industry.

The original triangle was published by Heinrich (1931). Heinrich was an insurance industry researcher who reviewed thousands of workplace accidents in order to determine their likelihood of occurrence to support actuarial calculations. The triangle has been redeveloped over time and expanded by others to add further layers of unsafe behaviours (Figure 5.1).

It is indicative of the development of safety thinking that a model developed almost a century ago remains an active, and indeed heavily relied upon, part of our current approach to safety management. If the same rate of progress were applied in other areas, we would still be starting cars using handles and have a thriving iron lung industry for polio sufferers. This is not to say that there is not development of ideas happening in the safety world – in industry and in academia – but that these have not been widely transferred into practice. Indeed, many have challenged the application of the triangle, yet it stubbornly refuses to go away.

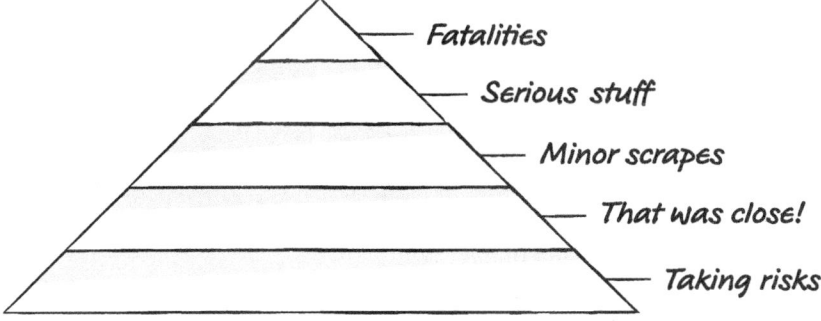

Figure 5.1 The accident triangle.

What Heinrich showed (among other things) is that we should look at accidents and their causes, rather than look at what injuries occurred. Thus we could improve injury statistics by looking for precursors rather than waiting for the accidents to investigate them. This was groundbreaking at the time and a great step forward in thinking, effectively forming the basis for modern leading indicator programmes. However, he attached some numbers to the model and that is where people fixated. Numbers have an almost magical quality in their ability to focus the mind. Any auditor will tell you that providing a score to accompany an audit or review has the effect of focusing all attention on how to get the numbers up, rather than carefully considering the underlying issues raised.

Heinrich's work should have been a footnote in the development of safety theory. A useful basis on which to build – aspects of it are incorporated into the 'Swiss Cheese' model and root cause analysis, for example – but once developed, left behind in its proper historical context. However, the triangle has a beguiling simplicity that allows it to remain in the textbooks. But this simplicity is superficial. Properly understood and considered, the triangle does not give the information that it is credited with, because not only is the triangle the most widely used tool in safety, it is also the most widely misinterpreted and misapplied.

There are many reasons for challenging Heinrich's work, including the lack of surviving data to corroborate the findings; the process for determining accident causes at a time when investigation techniques were poor (if used at all) and the fact that follow up was, in most cases, long after the accident. A number of other people have done so successfully in various papers and articles, most notably Manuele (2002). The intent here, though, is not to challenge the findings, so much as their application today.

The theory attached to the triangle is presented as follows:

- For every fatality there is a larger number of serious injuries.
- For each serious injury there are still more minor injuries.
- For each minor injury there are even more near-miss incidents.
- For each near miss, there are many unsafe behaviours exhibited.
- If we cut out the unsafe behaviours, then the near misses, injuries and fatalities will show a corresponding reduction.

The apparent genius of this approach is that it is *proactive*. It focuses on reduction by removing causal factors *before they develop into harm*.

It is easy to see why this approach gains such widespread approval. The holy grail of safety is the avoidance of harm. It is not acceptable to learn from our mistakes if those mistakes are fatal, so if we can develop methods for avoidance, it must be a good thing.

So far, so good. When individual incidents are investigated and analysed they may support this concept. In many cases, investigations will find early indicators of the failures that, if acted upon, would have prevented the worst-case consequences being realised. However, the detailed understanding of incidents comes from investigations which are, by definition, carried out after the incident has occurred. This misses the proactive point and so is replaced in most workplaces by the triangular fallacy. This has two key components:

1   confusion of correlation and causality; and
2   attempting statistical analysis.

Correlation and causality

This is the principal component of the fallacy. The confusion between correlation and causality is something that pervades much of our society, even rearing its head in research and academic circles where there is absolutely no excuse for it.

Human beings like to search for patterns. There are good evolutionary reasons for this and the brain is programmed to do so. The person who interprets a series of shadows as a predator and runs away can expect to survive longer than one who doesn't notice the pattern and hangs around. The majority of the time, the shadows will probably be benign, but when it counts it is important to be up and running.

So, when people see a series of numbers, they look for patterns and try to interpret them. Often, this is done incorrectly. A correlation occurs when two different numbers or facts can be linked together in some way. Causation occurs when one fact leads to the other.

> The majority of business leaders are degree qualified – correlation.
> Punching a wall hurts the hand – causation.

The majority of business leaders have degrees, but having a degree does not necessarily lead to becoming a business leader. There is a correlation here, but not causation. This can be clearly evidenced by the facts that there are many degree-qualified people who do not become business leaders and there are many (and many very prominent) business leaders who were college dropouts, or who never went beyond high school.

In the second case, there is clear causation. Punching the wall hurts the hand. It is not the only way to hurt the hand, but it can be shown that pain always follows the punch – certain and repeatable. There is causation.

The confusion occurs when someone applies a pattern or justification to the correlation, so attempting to extend it into causation. It is particularly tempting to think that degrees lead to business leadership because there is a certain compelling logic supporting the majority correlation. Degrees imply many of the intellectual skills required to make it in business. So it is easy to draw the conclusion that having a degree makes business success more likely. And when viewed across a large sample of people, this is true. But when looked at on an individual level it can quite clearly be seen that having a degree does not cause success in business. We see the pattern and ignore the instances that don't support it; all those people who have a degree but no job and all those successful non-degree-qualified businesspeople.

Many well-meaning parents may suffer from this when they encourage their children to study for a degree to become a business leader. Where this type of logic is involved and the correlation is broadly supported, the confusion is more likely to occur and be better rationalised.

In a similar way, the element of chance in outcomes is generally underappreciated when looking for causes. After an event, there is a tendency to attempt to rationalise what happened and look for reasons to explain it when, in fact, chance played a significant role. This is discussed in great detail by Taleb (2005). A more accessible account of rationalising the role of chance is given in Appendix 2 as a summary of a horse-racing betting trick undertaken by Derren Brown, the psychologist and illusionist, where he persuaded a woman to place her life savings on the outcome of a race based purely on the pattern observed. As a result of our need to find reasons for things, we clutch at the correlation and cling on to it.

Sometimes two sets of correlated data may be completely unrelated – for example, the divorce rate in the US state of Maine apparently correlates closely to per capita consumption of margarine. In other cases, there may be a third factor that connects the two data sets. Consumption of ice cream has been shown as correlating to drowning deaths. Clearly one does not cause the other, but a third factor – hot summer days increasing ice cream sales and swimming activities – may well underpin both.

In the triangular fallacy, the confusion arises because the triangle is viewed as a whole and not as a collation of discrete events. There is a correlation between the different levels of the triangle; however, there is not direct causation between them. While an individual event may have causal links to other individual events above or below it in the triangle, as a whole this is not the case. Nor does that causal link necessarily follow through all levels.

This assumption of causation throughout the triangle leads to a corresponding position that any action taken at the lower level of the triangle has an impact higher up. Focus is, therefore, placed on the bottom layers; if we can reduce the unsafe behaviours or near misses, the rest looks after itself. This means that a lot of effort goes into collecting data on lower-tier events and developing prevention strategies with much self-congratulation when first aid injuries reduce by 10 per cent. However, look more carefully through a causation lens. How many of those near misses could genuinely have led to a fatality, or even a serious injury? In many industries, a quick analysis of injuries will show cuts or bruises to hands or musculo-skeletal injuries as the most prevalent cases. What proportion of these injuries can lead to fatalities? In the majority of cases first aid or minor medical treatment cases have reached their full (or almost full) potential. There is no causal link to the upper levels of the triangle.

Despite this, I have heard many claims by safety people that they have saved lives based on the number of minor injuries avoided.

Excellent work by Phillip Byard (2009) has shown that the principal agencies that cause minor injuries are *significantly different* to those that cause fatalities and life-altering injuries. The only possible conclusion to be drawn from this is that effort expended reducing risks in the lower part of the triangle plays a very limited role in helping reduce more significant injuries and fatalities.

Where the safety professional, the workplace systems and management attention are all focused across the board on near-miss prevention, the vast majority of effort, resource, cost and time is being spent on incidents that will never escalate to anything of significance. This leaves only a small proportion of expended effort stopping the truly dangerous incidents from occurring. This aligns with the general findings of several studies that have shown that the gradual reductions in total injuries over the last 20 years or so have not been matched with a similar reduction in serious injuries and fatalities.

However persuasive the triangle correlation is, stepping back and using a little critical thinking paints a more realistic picture. Ironically, the safety professionals miss this while the workforce, those they are supposed to be protecting, understand it very well. Their response is often one of incredulity that they are filling in reams of paperwork and wearing protective equipment to avoid a minor cut or abrasion while serious hazards are not given the attention they deserve. A sure-fire way to lose the support of the workforce. This is the single most frequent complaint about safety that I hear when talking to workers.

At best, the triangle could possibly be redrafted as a stepped version with only the relevant causal factors making it to the next level (see Figure 5.2).

But even this is flawed, as there are many instances of fatalities where there is no minor version. While it is possible for unsafe behaviours and near misses

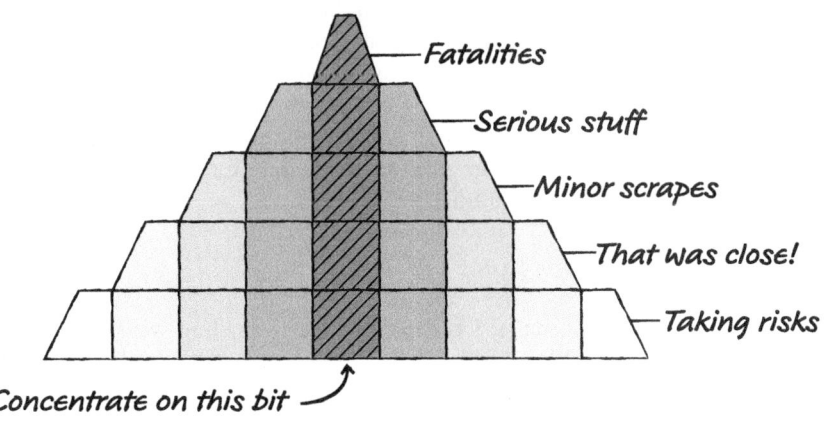

Figure 5.2 A risk-adjusted accident triangle.

to be precursors of a catastrophic explosion, for example, for the most part there are no first aid injuries caused by explosions (except on the very periphery with other people more seriously injured closer to the blast, which is not much of a proactive prevention statistic). The failure jumps straight from near miss to fatality with no intermediate steps. There will be indications if you know where to look, but that is an entirely separate discussion around process safety indicators.

One of the most illustrative techniques for demonstrating appropriateness of a safety approach is to translate it into a different function (usually financial) and seeing how it stacks up. Imagine in this instance that we are trying to prevent fraud within a business.

In triangular management terms, we would look at the total number of frauds. If we stop them when they're small, we won't get big ones. There are many more minor frauds than major ones, so we put our focus there in order to easily show reductions in events (getting the numbers down). We put a lot of time and effort into making sure that expense claims don't include drinks with dinner that can't be claimed under corporate policy. And setting up and monitoring a gift register with anything over $50 in value on it. At the end of the year, we can show our total number of fraud incidents has dropped from 50 to five and everyone is happy. Of course, with all this busyness going on, we didn't have time to catch up with the one person who was approving their own purchase requisitions as part of a large capital project. As we didn't measure severity of fraud, just numbers, we didn't see that the total cost of fraud went up by several hundred thousand.

Of course, what we should have done is reviewed the business and determined where the highest risk of material fraud lay and put some effort into closing loopholes in the purchasing system and adding a layer of review and double-checking. It's the same in safety. We should review what we do, recognise where the significant potential lies for severe injuries and fatalities and put most of our effort there. We may miss the odd cut finger, but everyone will have a greater chance of going home alive.

The use of the triangular fallacy to misdirect effort away from incidents with significant potential is one of the biggest problems facing the effective management of safety. Businesses have limited resources and ensuring they focus on the right areas is fundamental. In this light, the triangle (or at least its interpretation and use) is therefore not only flawed, but *actively working against the prevention of serious injury and death.*

Statistics

The inappropriate use of statistics is the second major failing with application of the triangle theory. This is not so widespread or damaging as the causation confusion, but is nevertheless symptomatic of sloppy thinking. We will discuss the general inability to deal with statistics in a later chapter, but the specific issue here is one of blind application of numbers.

The statistics in the original triangle were based on Heinrich's reviews of a variety of industries in the United States in the 1920s. It should be quite apparent that the triangle will be a different shape for different industries. High-risk industries (say, steeplejacks) can expect a much taller and thinner triangle, given the much higher potential for fatalities. On the other hand, a call centre may expect a very low and flat triangle with a very high proportion of injuries to deaths.

It is equally apparent that any collection of such statistics only gives a point-in-time snapshot of the industry. The same industry over a different period would produce different figures, with the degree of variation in inverse proportion to the size of the sample.

It is clear from this that only a limited amount of predictive information can be gained from the numbers in the triangle. Companies may pool data, or look at industry-wide data to get greater consistency with a larger sample size, but this will make the data less specific to their operations. Conversely, using data only for their own operations will lead to variation year on year. Despite such obvious shortcomings in the data, there are companies routinely basing their future safety programmes on the numbers in the triangle and their ratios. Statements are heard such as, *'Our ratio is 10:1, but we have only seen 3 serious injuries this year despite 75 first aid cases, so we're overdue one'*, or *'We must look for more unsafe behaviours because we aren't getting enough for the number of injuries.'* In either case, the active focus is being developed for the future based on entirely unsupported data assumptions. It's almost as if it would be preferable to have more injuries if the numbers lined up with the model.

Now, it may be that taken over very large populations covering a number of industries, ratios can be found that are consistent – but that means nothing in the context of the individual business attempting to manage its own operations. Try suggesting to your management team that salaries should be based on the national average wage rather than careful consideration of industry- and location-specific data.

At least some organisations base the ratios on their own incident history by monitoring their own ratios to gain an understanding of performance. However, I have heard a corporate Health and Safety Manager (working at a national level in one of the country's larger organisations) state that the company was overdue an injury on the basis of the ratios in Heinrich's *original* triangle. This is the absolute extreme, where policy decisions were being made based on data from a number of companies in a different country and different industries that were collected almost 100 years ago.

If nothing else, what does it say about the quality of our management of more significant events over the last century if the same proportion of minor injuries ends up as fatalities? Surely, at the very least, improved emergency response would have changed that?

One of the key things I learned from many years of quantified modelling of accident modes is that the reality rarely fits the model and if they're different, surprising as it may be to some, reality wins. As is often said, all models are wrong but some are useful.

I have taken to challenging the triangular fallacy wherever I see it and have yet to hear a convincing argument in favour of it when applied in this way. Its role in driving the identification of accident precursors was hugely important in the development of safety practices, and as a visual representation of the potential escalation of the severity of a single incident it may also have a place in training and awareness. Unfortunately, it has been so tainted now that it would be better to remove it altogether and consign it to history.

Key points

From this...	...to this
• All injury prevention supports reduction in major events.	• All injuries are not created equal. Stop treating them as if they are.
• Accident prevention is focused on near misses and at–risk behaviours across the board.	• High-risk hazards are clearly understood and the focus is on identifying and managing their specific precursors.
• Statistical correlation is assumed between different injury levels.	• The focus is on causal relationships and potential harm from specific scenarios.
• The accident triangle is used to guide intervention activities.	• The accident triangle is thrown out.

References

Byard, Phillip (2009) *Is Zero Harm Killing Our People?*, Intersafe.
Heinrich, H.W. (1931) *Industrial Accident Prevention: A Scientific Approach*, McGraw Hill.
Manuele, Fred A. (2002) *Heinrich Revisited: Truisms or Myths*, National Safety Council.
Taleb, Nassim Nicholas (2005) *Fooled by Randomness*, Random House Trade Publications.

6   The priority confusion

'Safety is our number one priority.'

Is it?

Really?

This is a statement that is trotted out so often by managers that it has become a tired and worn cliché. I am normally a strong defender of the cliché, since the reason they are so well used is because they are true and highly applicable in many circumstances (this will also be my defence for any you may find elsewhere in this book). While their frequent repetition may be unpalatable to authors, journalists and others interested in language for its own sake, in factual terms they can be seen as the succinctly phrased culmination of years of practical experience.

In this case, however, there is one small, but crucial, difference. It's nonsense.

There are other variations on the theme. 'Safety First' is the most common one, shouting stridently from posters around the world. But others abound, usually some version of 'Safety is our number one priority.'

The safety profession has a bad habit of speaking in absolutes. Black and white. Unfortunately, in a world of complexity and greyness, there is little room for this type of language. It sets us up to fail to deliver against our promises.

The main problem is that safety is not the number one priority and nor should it be. There are always competing priorities and in business the number one priority is generally to make money, or at least to stay in business. I'm not suggesting that it's okay to allow people to be hurt in pursuit of cash, just that we need to be a little more practical as to how we go about it.

Here is a test to understand where safety really sits as a priority in an organisation. There are three questions:

1   'If I could give you a tool that is guaranteed to reduce your injury rate by 90 per cent, would you be interested in implementing it in the business?'
2   'If it would cost you all of your profits for the next five years would you still be interested in it?'
3   'Are you positive safety is your number one priority?'

The answers are usually some variation on 'yes' to the first, some mumbling in response to the second and 'it's more complicated than that' for the last. And of course it is more complicated than that. This is a ridiculous extreme to make the point. But that point is that if you are going to go around talking in absolutes such as 'nothing is more important than safety' or 'safety is always number one', then you had better be prepared to defend that position against an absolute counter-argument.

Anyone who still says yes to all three is either an outright liar, hiding behind the hypothetical nature of the question or a genuine safety fanatic who will have their view overturned by the board pretty sharply.

A more realistic scenario than our hypothetical questions may be that a business is tendering for a large job. They have a high chance of success and it will secure a number of staff positions for two to three years. Without it, those people may have to be made redundant. It is a government contract and so the rules of engagement are very strict. The tender documentation must be delivered in hard copy to the department by 5 p.m. today. At the last minute, the company network goes down (of course) and there is a delay in developing the documentation. The package is pulled together, but is only ready at 4.30 p.m. It is now rush hour and the only way to get the package out in time is to use a motorcycle for delivery. Travelling under pressure to meet a deadline in heavy traffic on a motorcycle is very high risk. You may even have a (safety first) policy not to use motorcycle couriers because of this risk. Would you approve the journey?

There is no right or wrong answer. It may depend on the distance of travel, the speed of rush-hour traffic, the experience of the rider, the weather, the time of year. The point is that the question is not simple. In addition, there is the microscope of hindsight to make matters worse. If you decide 'no' and it turns out the traffic was particularly light that evening, you need to justify that to the staff you are about to make redundant. If you decide 'yes' and there is an accident, you may need to explain it to the judge (or at least the CEO).

There are two types of priority confusers – the Well Intentioner and the Lip Servicer. We'll deal with the Lip Servicer first as they are more widely recognised. We will meet both types again later when we talk about zero-harm goals.

Lip Servicer

This person is paying lip service to the safety concept. They don't believe in it, they don't support it, but they do know in this day and age that they have to say it because that is the done thing.

These are also the same people who may say that people are their number one asset, and then cut the training budget at the first sight of financial trouble; who 'only hire the best' yet peg their salaries to industry medians; who have an open-door policy but react angrily to hearing bad news.

In short, they are lying and, sooner or later, everyone will find out when their actions are inconsistent with their words.

Because these are obvious and everyone recognises the type, I'm not going to dwell on them.

Well Intentioner

The good intentions of this person are clear. They genuinely believe in what they are saying; they are simply misguided. This is where the critical thinking step has not happened and unintended consequences can arise as a result. They are sometimes a manager being badly advised by their safety team, but safety professionals can be the biggest culprits of the well-intentioned priority confusion.

Running a business is complex and multifaceted. There are always many different priorities to consider when making decisions. It helps decision-making processes to have a clear understanding of what your values and principles are, so that you can make a choice that is right for you and does not violate your values. However, priorities can still shift within a principle-based framework. If you have a single stated priority that overrides all others, it is inevitable that at some point there will be a time when this is not appropriate for the overall good of the business.

As an aside, there is a slight variation on this that is becoming more common among safety professionals. They say that 'Safety is not a priority, because priorities change. Safety is a value as that is constant.' This is true. Safety should be a core value, but beware the underlying message which is, 'We recognise that priorities change, but we're not prepared to accept that safety isn't number one, so we'll call it something else instead and then we can still go on about it being more important than everything else.'

Taking an extreme position again, if safety were the most important thing, we would never go to work. That way the risk of a work-related injury is removed. This is, of course, ludicrous, but it does demonstrate that every decision we make in going to work has a risk trade-off (and actually, it may not be true – statistically, it may be safer to force everyone to stay at work than go home, although in either direction the journey is probably the biggest risk for most of us). And this risk trade-off does not stop on arrival at the office or factory. It continues all day – should I climb the ladder or build a scaffold? Should I replace the fuse, with a risk of electrocution, or purchase a new appliance? In most cases, this trade-off is simple because the risk is very small and the cost of risk avoidance comparatively high.

But what about when the balance becomes more even and the disparity smaller? What if there is a large risk, but also a large cost? By committing to an overriding priority you run the risk of either making a decision that is not in the best interests of the business in order to remain consistent, or of making the right decision but being accused of hypocrisy. Neither is a good look.

Consider the following scenario, which may well have played out in a number of workplaces over the last few years. A business operates a fleet of vehicles. One year the manufacturer brings out a new model that has a new safety feature – electronic stability control. This makes the vehicle safer, and a member of staff asks if they should have the newer model. Depending on the age of the existing fleet and the cost of terminating the current lease arrangements, this may cost upwards of $10,000 per vehicle. For 50 vehicles, that's $0.5 million to replace a fleet that yesterday was full of the safest vehicles on the road (because we're a safety conscious company). Can you say no to this if you have spent the last few months on a priority confusion crusade telling everyone how safety is more important than anything else? But can you say yes to the cost and look the majority shareholder in the eye?

Had you not been on that crusade you can explain and justify your decision. It may not be agreed with, but at least it can be respected. Otherwise, you need to either come up with the cash (hoping that there isn't another innovation from a different vehicle manufacturer the following month), or contradict yourself. There is nothing a workforce spots faster than inconsistency. This is labelled as hypocrisy and will undermine all of your future exhortations for safe work practices, providing another way for the workforce to disengage. This is particularly galling for the Well Intentioner as they are genuinely trying their best.

Additionally, there are disruptive individuals out there who will use the priority confusion to their advantage by playing a safety card where none really exists: 'We can't use those vehicles because they're not the safest and you said that safety was our number one priority.'

There are times when the correct option is not the safest one. People largely understand this and agree with it – hence the societal backlash with allegations of a 'nanny state' when over-the-top safety rules and regulations are applied. Risk is a natural part of life and is generally accepted (and even appreciated) as such. It is only when the muddled thinking of safety bureaucrats (let's call them safetycrats) is applied that common sense is rejected. The safetycrat is only interested in removal of risk, not management of it, mitigation of it or even (heaven forbid) acceptance of it.

> When safety is our priority, we live our lives being very, very careful, and we wind up having no lives.
>
> (Byron Katie (www.thework.com))

There is a common misconception in the general public that the law and policy makers are responsible for the safetycrats' behaviour. In fact, in many jurisdictions, the legal framework very specifically includes terms such as 'so far as is reasonably practicable', 'all practicable steps' or other similar versions. These take account of the fact that there are other priorities to balance and that if a safety improvement has costs that are disproportionate to its benefits, it may not be necessary to implement it. And, let's face it, government legal drafting departments are hardly filled with the most flexible and pragmatic of folk. When we, as a profession, are more black and white than they are, we really need to take a long, hard look at ourselves.

The UK Health and Safety Executive (UKHSE) website has a section devoted to myth busting, where ridiculous activities taken in the name of health and safety are debunked.

How did this position arise for the Well Intentioner? Simply bad advice. The safety professional is speaking through the mouth of the manager and usually also has good intentions. It is a common finding that emphasis of production over safety is a causal factor in accidents. It is important, therefore, that it be emphasised that safety take precedence over production, and it is a small step from 'We must take time to do our job safely' to the full-blown priority confusion.

As we have seen, though, there are many degrees of risk, and stopping production for a very minor risk may be inappropriate. Yes, safety should be considered as a high priority, and the more significant the likely consequence the more weighting the safety implications should receive in the decision. In searching for a slogan, 'Safety is our number one priority' certainly seems more appealing than 'Safety is one of a number of, occasionally competing, priorities that we take into account when making decisions and, while we will weight safety at a level commensurate with the perceived risk, it will not necessarily override all other considerations.' This certainly wouldn't look great on a poster.

While the priority statement is well-intentioned, it is the unintended consequence that needs to be taken into account. To maintain worker engagement it is vital to maintain congruence between words and actions – to walk the talk. Any failure in this area reduces credibility and can undermine much broader safety approaches, initiatives and culture. By using the priority confusion phrase, incongruence is almost inevitable.

This is a subtle impact, but can be very significant – absolutism is difficult to pull away from and should be avoided in any complex environment. A key factor in the priority confusion is an assumption that any injury is unacceptable. Indeed, this is a commonly quoted variant on the absolutism of 'Safety first'. However, the reality for most people is that injuries are acceptable – but only so far and only in accordance with the prize at the end of the activity being undertaken. After all, nobody, not even the most zealous safety crusader, would suggest that a light bruise from walking into a desk is unacceptable. There is no call for all desks to have rounded edges or cushioning installed. We are prepared to accept the injury due to a combination of its triviality and the ridiculously disproportionate cost of prevention. What safetycrats fail to understand is that the point which they see as the minimum acceptable level may not be shared by others and, indeed, is not constant. We will talk a little more about risk perception and risk acceptability later on.

What is safety anyway?

A good question to ask the priority confuser is what safety means. As discussed earlier, we all talk about safety a lot, but it actually means different things to different people. Having this conversation can often help people to see the fallacy in the 'Safety first' slogan. In defining safety, it becomes necessary to think about shades of grey, because absolutism is difficult to justify in an explanatory way. It is so easy to point out that activities can be undertaken unsafely without injury and that some degree of minor injury is acceptable in context that the absolute position quickly fails.

The best working definition of safety I have used is 'freedom from *unacceptable* risk'. There are more philosophical definitions than this that are based on safety as a positive factor in supporting productive work. Indeed, Hollnagel (2014) specifically deconstructs this definition as inadequate. But it forces the discussion of what is acceptable, and this brings with it the many different perceptions and tolerances that are in place among different people. For the majority of workplaces it is a good intermediate step to at least get them talking about grey areas and considering it as a first movement away from black and white.

Messaging about the importance of safety is necessary, so what are the alternatives? The message needs to be part of a broader overall one about the company's values and approach, and so it is difficult to make suggestions in isolation, but the following suggest emphasis on safety without absolutism committing you to an outcome:

- Safety is a core value.
- We build safety into all that we do.
- We always think about safety before progressing.

While there is still scope for disagreements and arguments within an overall framework because safety is a relative concept, in this way a position can always be established, discussed and, if necessary, defended while retaining credibility. Absolutism should be avoided. It is the foundation of propaganda, of dictatorship and repression (in fact, at the time of writing the built-in thesaurus has absolutism as a synonym for tyranny and an antonym of democracy). It is not the basis of an open, learning and sharing culture that is required to best develop safe working practices.

Key points

From this…

- Safety is discussed in absolutes where it *must* come first.

- Injuries are considered unacceptable.

- Statements are made with little thought to their impact.

- Safety is black and white.

…to this

- Ambiguity and context are recognised as inevitable and included in safety discussions.

- It is recognised that nothing is ever risk-free and the downside risk is balanced against what is trying to be achieved.

- Statements are made with an eye to the future and do not commit absolutely to an outcome when future factors are unknown.

- Safety is ambiguous, with shades of grey.

Reference

Hollnagel, Erik (2014) *Safety-I and Safety-II: The Past and Future of Safety Management,* Ashgate.

7 The benefit façade

What is the cost of an accident?

This is a question often asked during training courses on accident investigation. It is a valid question and it is useful in an investigation to state the actual or potential cost as an emphasis for the importance of preventing recurrence.

There are a number of facets to the cost incurred. There is, first and foremost, the human cost – the individual or individuals hurt, their families, friends, colleagues or managers. The more severe the accident, the higher the cost in human terms, but also the further that cost extends beyond the people immediately involved.

Then there is the financial cost, which can be felt via a number of channels from the accident to the annual report. This includes operational down time in the immediate aftermath, the cost of the investigation, replacement of damaged equipment, cover for injured staff, compensation payments, fines, increased insurance premiums, increased government levies (in certain jurisdictions), etc.

Finally, there is a reputational cost to the business. Although, in reality, this is simply a different route to a financial cost as it results in potential loss of earnings, inability to attract good staff (who wants to work for an operation that kills people?), future investment partners and so on.

Following discussion of the cost (preferably with a worked example including jokes about the hourly rate of lawyers), the conclusion is drawn that it is worth spending time, effort and money to properly investigate to find and eliminate root causes that may cause this and other, similar incidents from occurring again and incurring said cost.

So far, so good. You can't argue with the logic, and the beauty of the conversation is that it includes cold hard cash costs and that gets the attention of senior management. As everyone knows (apparently), you have to talk to management in profit terms because they don't understand anything else. However, as we know from Chapter 2, this is not actually true in the majority of cases and we will discuss more later about talking to the senior executives and directors in the business.

Within this discussion of costs and the associated benefits of proper investigation, there is a seed of a broader concept. Good safety practices are shown here to provide cost avoidance for the business. Not quite as good as an actual income stream, but pretty close if the costs being avoided have a reasonable degree of certainty. Traditionally, most safety professionals complain, with some justification, of the reluctance to spend on safety because it is seen as a cost, an additional overhead that cuts into margins. And, to make matters worse, for most people it is tied in with activities that slow down production as well. Using our newly determined accident costs, we can show that although there is an initial cost, it will save more than it costs in the long run by avoiding future spend. There remain some difficulties in making the case because an uncertain cost in the future is given less weight than a definite cost now, but we can see that safety becomes a valid insurance policy. Pay a small amount up-front – avoid large costs later.

This is still all good, until the next step is taken. If this is true for investigation, what about other areas of safety? Are there other benefits we can sell to get traction for our safety implementation? Cue numerous studies that identify the benefits of a safe workplace. Those organisations that have better safety records also demonstrate better productivity, greater workforce engagement, better product quality and, ultimately, better financial performance.

This is the benefit façade. It states simply that good safety is good business. It is well meaning and in many respects it is harmless. The façade is less a dangerous myth and more a missed opportunity to get even better performance. Although it can be costly when someone tries to sell you the latest silver bullet or snake oil safety improvement system based on a sales pitch that good safety improves the bottom line of the business.

The reports into the benefits of safe workplaces are further examples of the correlation versus causation problem discussed previously. They have all the hallmarks of a close correlation and persuasive logical links between the elements under consideration. Note that the following discussion centres on how these reports are broadly used in the industry. Individual researchers and their specific findings may be more circumspect in their claims and more accurate in their interpretation of the data (and I believe that in the area of occupational health, there is some more validity in the arguments than in occupational safety). Essentially, the argument goes something like this:

Good safety means people feel more valued because their employer cares.
When people feel more valued, they feel more engaged at work.
When staff are more engaged, absenteeism and presenteeism★ fall and productivity increases.
With increased productivity comes increased profitability.
QED – good safety means more profits.

★*Presenteeism* – the concept of being at work, but not actually doing anything worthwhile the majority of the time, usually as a result of feeling disengaged or disenfranchised

In order to test the hypothesis, organisations are studied and those identified that have good safety systems in place according to some suitable measure such as external certification to a safety standard, or possibly purely based on accident rates are also shown to outperform their peers in business metrics. Lo and behold, our assumption is proven correct. Good safety is good business and if you invest in safety systems, your profits will go up. Although it tends not to be stated quite as baldly as this, because such a statement invites some searching questions. More likely is a statement that companies with good safety systems also demonstrate better business performance, which is somewhat more defendable.

Actually, I don't have a great problem with this per se. What concerns me is the *missed opportunity* that proper understanding of this can bring.

What else could be measured that could correlate to good business performance? What about HR processes and people management? These include how well people are recruited and inducted, how well they are remunerated and how much they are listened to and empowered. These things all make people feel more valued, which in turn makes them more engaged – sound familiar? So maybe good HR processes provide business value and, no doubt, there are HR periodicals out there making that same point and providing surveys that justify it. And they are right.

What about proactive equipment maintenance? In a manufacturing or processing facility, particularly one which is either capital-intensive or highly sensitive to downtime, good, well-thought-out preventative maintenance can have significant uptime benefits later on. This makes the up-front cost a worthwhile investment. Good design and commissioning minimise infant mortality rates in new installations (infant mortality here refers to failures early in the life of new equipment – we're not looking at maternity units, although this probably applies there equally). It is very clear from research that the best-performing companies have top-notch maintenance practices that are nurtured in the design phase (this is an interesting and important issue in its own right that I don't have the expertise to go into). Perhaps this is the key to good business performance?

I have no doubt that there are quality specialists, customer service gurus, accounting savants, marketing experts and many other disciplines out in industry all selling the same message that investment in their area of expertise will result in stellar performance.

And here is the key – *they are all correct*. Let's face it; if you can't demonstrate a link between your department's performance and the overall business performance, you really ought to question what you're doing there.

In my experience when I have worked with good organisations in safety, they are good in all these other facets as well. There may be variations across the business, but in general terms a good-performing, highly profitable company will be good at safety, they will also produce a good-quality product and have good marketing, good customer service and good relationships through their supply chain and so on and so on.

The missed opportunity is recognition of the fact that all these aspects are interlinked. Safe production is reliable production is cost-effective production. Trying to impose a good safety system onto an otherwise dysfunctional organisation may deliver some marginal improvement to overall performance, but will ultimately fail in any meaningful way to provide broader business benefits.

To develop any significant system improvement, management involvement is essential to sponsor and drive the change. This is not easy to come by, especially if the organisation is in the not-so-good category. If the safety professional can achieve management support in this environment it is absolutely vital that success can be demonstrated. If not, credibility will be lost, support will be withdrawn and the chance will have gone. It is unlikely to come back. So, simply stating that improved safety will be good for business is not enough. Because it won't work unless the underlying performance issues of the business are addressed first.

Safety professionals must think more holistically. They must recognise that there are other aspects to managing a business that must be improved hand in hand with safety. They must get this message across to management. If this is not presented as a united front across all systems and functions, safety will be targeted for improvement in isolation and failure will ensue. It is imperative that this united front is established at the very outset. Delivering improvements inevitably requires time, effort, resource and money to implement. When management has approved funding for safety improvements they will be likely to see them as exhausting their general improvement budget. Requests for expansion to cover other functions will fall on deaf ears. But if the extent of the requirement and its associated benefit can be made clear at the outset, there is a chance for success. Managing expectations is a key factor – and this will be a long and slow journey.

Of course, the problem of the benefits façade is exacerbated hugely by the priority confusion discussed previously. The chances of this holistic approach being taken are low if the belief is held that safety is the most important part of the business. Even when some overlap with other functions is recognised, the obsession with safety will tend towards treating it with a higher priority at the expense of other areas, eventually leading to the same end result.

But it is very easy to identify the links to the rest of the business:

- Good safety requires good, up-to-date, properly approved procedures – that is a QA function.
- Good safety requires good-quality and well-maintained equipment – that is a maintenance function.
- Good safety requires well-designed equipment with appropriate interlocks, guards and inherent safety design features – that is an engineering function.
- Good safety requires correctly specified repair materials to be available – that is a procurement function.

- Good safety requires staff to be motivated, adequately resourced and well managed to avoid fatigue and distractions – that is a line management function.
- And so on ad infinitum.

Ultimately, the overall performance of a business is more strongly linked to its culture than anything else. As mentioned earlier, culture is a fascinating subject that deserves several books in its own right, but it is also a widely abused term by would-be experts. There is much discussion about the importance of a positive safety culture in determining good safety performance. But a good safety culture cannot exist in isolation from the rest of the business. Excellence, and the culture that supports it, is not a pick-and-choose exercise. You can't be excellent at safety and then turn around and be sloppy in other areas. Excellence is a mind-set, a philosophy, a habit.

> We are what we repeatedly do. Excellence, then, is not an act, but a habit.
>
> (Aristotle)

It is easy to tell which of the tradesmen on a building site is the most likely to be the safest. The one with the tidiest work area, the cleanest and best looked-after tools. The one who takes most care and pride in the work that is produced.

As it is with individuals, so it is with companies. A company with a good culture will extend that good culture into safety, because they are in the habit of listening to their staff, of taking the time to do things the right way, of producing quality outputs, of optimising their processes, of learning and improving. A good safety culture, or good safety performance, does not lead to good business output. Rather, the underlying aspects of a good safety culture – communication, openness, honesty, transparency, valuing learning, continuous improvement, care for people – are also the underlying aspects of good quality, good customer care, good workforce management, good productivity management and so on.

By focusing purely on safety, we are both missing an opportunity and increasing the likelihood that the improvement will fail. The genuine benefit is to be gained from recognising this connection and working on the business culture as a whole. The rewards will then be reaped not as a short-term boost to safety, but as a genuine business improvement that can be sustainably achieved.

To support this process, safety managers need to become more conversant with the rest of the business processes. Not only because it will help in understanding these inter-dependencies, but because they need to be more influential within the business to make the case. Safety professionals are rarely included within the executive team. Executives need to be generalists. There are many and varied aspects to running a business and those in charge need to be *au fait* with finance, people, management, leadership, strategy, marketing, PR, regulations, customer relationship management and much more. We will discuss some of these aspects later when covering the communication gap, but until safety professionals understand enough about the rest of the business to bring genuine value to the leadership table, they will not routinely form a part of the executive team. Until they do so, they cannot bring to bear as much influence as they might.

Key points

From this...

- Safety is treated in isolation and separated out from other business functions.

- Safety improvements are made by working on the safety system.
- Good safety leads to good business.

- Pockets of good practice are seen in the business.

...to this

- The interrelationship with other business functions is fully understood and incorporated into safety management thinking.

- Safety improvements are made by first working on the business.
- Good safety performance and good performance elsewhere are underpinned by the same behaviours, disciplines and practices.
- A habit of excellence pervades the business.

8   The number crunch

A book about numbers doesn't obviously jump out as the most enthralling read (although this statement may be a bit hypocritical coming in a book about safety), but *The Tiger That Isn't* (Blastland and Dilnot 2007) is well worth reading. In it, the authors investigate the world of numbers, with a particular emphasis on media spin and distortion of numbers to allow conclusions to be drawn in error.

Numbers can be hugely persuasive. They are presented and received as fact, often with little underlying analysis to support the veracity of the data. Conclusions are drawn based on small data sets and wildly extrapolated to apply in all sorts of irrelevant areas. One of the principal inferences from the book is that most people are easily duped by numbers because they are statistically illiterate (I use this phrase rather than innumerate because the majority of people can add up).

Statistics are used widely in reporting safety performance. How many accidents have we had? What is our total recordable injury frequency rate? How do our numbers compare against others in our industry, or in other industries? In many cases, these so-called 'lagging' statistics (i.e. numbers after the injury has occurred) are the only information a company has to go on to make its broadest policy decisions around safety. Even major companies with sophisticated and extensive safety programmes still essentially only report lagging indicators in their annual reports.

But how well do we understand the numbers? How many safety professionals have any kind of higher-level qualification that is strongly numerically based? Or, for that matter, how many of our management team genuinely understand numbers?

I expect, if you're like I was, you are reading this and ascribing this failing to other people: 'I understand numbers and I'm not easily duped.' So here are a couple of questions for you.

Don't cheat and read ahead. Answers in a few paragraphs' time.

1   [Easy] How many eyes does the average person have?
2   [Harder] A disease affects 1 per cent of the population. The test for the disease is 90 per cent accurate. What is the likelihood a person diagnosed as suffering actually has the disease?

The point here is not to be clever, but to show how some so-called obvious assumptions can be wrong. Question 1 shows a little about the concept of average (or *mean* in statistics). We may benchmark against industry averages, but what does that mean (no pun intended)? The answer to the question shows that almost everybody has more than the average number of eyes. People tend to think of the average as being typical of most of the population. But actually that is often far from the truth. Single outliers can cause huge fluctuations in the average. This is especially true for data populations that are asymmetric, those that are constrained at one end – usually by a minimum – but not at the other.

The most obvious example of this is income. Income is constrained by a minimum of zero at one end of the scale but, theoretically at least, there is no maximum at the other end. Let's imagine you're at a meeting with eight colleagues when Bill Gates comes in the room. Gates has an income of, for argument's sake, $5 billion per year. The average salary in the room is now about $500 million. Is that representative of the participants? Not in any meeting room I've ever been in.

Injury rates or fatalities cannot drop below zero but can be very high, albeit with a practical upper limit based on total exposure. Safety statistics are, therefore, subject to the same skewed distribution with the same impact. This means that average industry rates are not necessarily a good measure. One bad performer raises the average disproportionately, which makes everybody else's performance look better.

What does having unusual outliers in the data set mean for industry benchmarking of performance? Let's take a selection of ten companies in a single industry that all report injury rates in the same way and publish their information through a confidential third party in the form of an industry average.

Nine companies have an injury rate of five per year, the tenth has had no accidents. The average rate is 4.5. The nine high companies worry that they are above the industry average and implement some practices to improve.

The following year, company 10 moves from zero to ten (it was a bad year). All the others remain at five (their improvements were not very effective). However, now the average is 5.5 and all the other companies are better than average. They pat themselves on the back and do more of their ineffective safety processes.

Of course, it would take a seriously statistically illiterate person to draw that conclusion, but given the unpredictability of benchmarking data (and the way the data is presented to management with little underlying detail), it is not so far-fetched. Now throw in the added complication of differing staff numbers, differing roles being worked across the population, different weather conditions in different locations and other factors, and the whole benchmarking process can become a waste of time.

For skewed data distributions such as this, a better tool to use is the *median* rather than the *mean*. The median is the point at which 50 per cent of the population lies above and 50 per cent below. This is used in politics and economics to discuss national salaries, for example, to counter the issue described above. However, there are many people who don't understand the difference between the two and will use them interchangeably, or will simply stick to using the mean. A quick Google search on 'industry average injury rates' and 'industry mean injury rates' returned a combined total of 66.7 million results (using both as the terms are synonymous). Change 'average/ mean' for 'median' and the result falls to 6.9 million. Hardly scientific, but it gives a feel for how often each is applied (although no clue as to whether they are applied correctly).

Question 2 is unlikely to be encountered in most workplaces, but statistics taken across entire industries or major multinationals may develop into something as complex as this. This is a medical question, but even doctors, as smart and well-educated a group as you could come across, have struggled to get it right. This shows that data management needs to be treated with caution. This second question demonstrates particularly well the danger of complex statistics. At first glance, it looks as if the likelihood of having the disease in this situation is high (90 per cent), but in fact it is low (<10 per cent). This is an enormous swing and can have a major impact on decision making. Not simply inaccurate, but actually the opposite of what is needed.

The answer to question 1 is slightly fewer than two. Most people have two eyes, but there are a few with one or none and nobody with three. The one/none population reduce the average to slightly below two.

The answer to question 2 is approximately 9 per cent. The disease affects 1 per cent of people. If 1,000 people are tested, 10 (1 per cent) will actually have the disease. Of these nine will be accurately tested because the test is 90 per cent accurate; one will be falsely cleared. However, of the remaining 990, 99 will receive false-positive tests. So 108 people will test positive in total, of whom only nine will really have the disease.

Consider the way data is presented to the executive and board within your organisation or your experience. Can you genuinely say it is carefully examined with all the underlying detail analysed? Are insightful questions asked around whether the conclusions drawn are accurate and realistic?

Statistical traps

One of the most popular (and useful) ways of presenting data is in the form of trends. It is no use knowing that we have injured seven people this month if there is no context around that – is seven better than last month, worse or the same? If different, is it significantly different, or only marginal? Is that difference part of an ongoing trend, or a single outlier? Providing trending data allows some of these questions to be more thoughtfully considered. This is frequently done using a '12-month rolling average' (or rolling average over some other period). This is primarily because we need a timeframe for the trend, but if we choose calendar or financial years, the year is halfway through before we start to see the trends. A 12-month rolling average takes the average of the immediately preceding 12 months and plots it on a graph so that there is always sufficient data to see a trend. This is a perfectly legitimate process, but, unsurprisingly, there are traps in the detail.

Take the data set for injuries presented in Table 8.1.

This shows a clear downward trend. What we are doing appears to be working (subject to the problems of using lagging indicators, which we'll talk about later). So we should continue to do more of the same and investigate other improvements (because we never stop looking for improvements). Shown on a graph, this should demonstrate our month-on-month improvement for all to see. Shouldn't it?

However, now imagine that the month before this data set, the previous December, we had a stellar month in which we had no injuries. In last month's information, this zero was part of our rolling average, lowering it. This month, however, it has dropped off our data set, so we no longer have the benefit of it in the average. The rolling average graph looks like Figure 8.1.

The way the data is displayed presents the exact opposite picture of what is happening. 'Something is wrong here. We've had a sudden spike. Let's change everything.' Without the careful analysis and questioning that needs to go with it, we end up in the same position as described in question 2 above – drawing the completely wrong conclusion.

Table 8.1 Monthly injury rates

Jan.	Feb.	Mar.	Apr.	May	Jun.	Jul.	Aug.	Sep.	Oct.	Nov.	Dec.
8	8	8	8	7	7	6	6	5	5	4	4

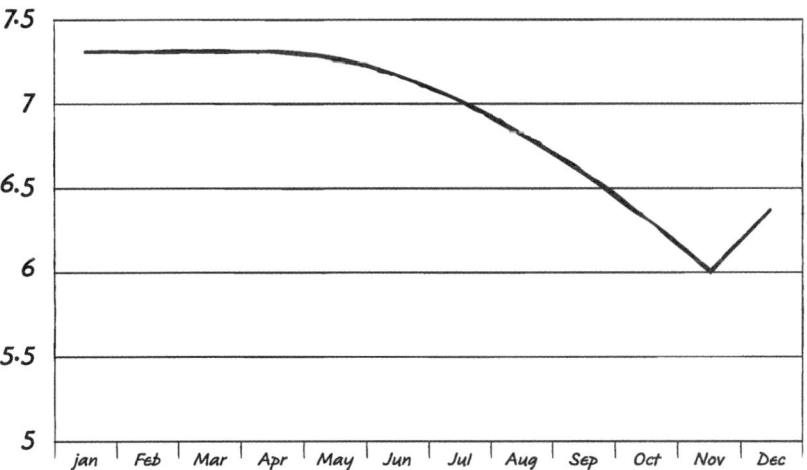

Figure 8.1 Twelve-month rolling average.

This leads us to the single biggest failing with safety statistics – drawing conclusions from insufficient information and basing our actions and strategies on them. Even if Figure 8.1 represented reality, it is a single data point that is inconsistent with a general trend. We should be very careful about knee-jerk reactions to it, because next month we could revert to type. This is particularly of concern at executive and board level where the sheer volume of information required to govern the business lends itself to short summary statements. Summary data should only be used as a starting point for enquiry – never as a decision-making tool in isolation. Such statements should always be accompanied by an explanatory narrative, even if very brief, to assist in understanding the underlying factors and preventing inappropriate responses.

In fact, single outliers are generally a problem if there is a temptation to act on them. This is also true for a few data points that look like they're a trend. On a construction project building a new gas-processing facility we had implemented a random drug-testing regime for the site. Once a month (or so) a random group of people from the site would be selected for the test. In the course of the first three tests, one individual was chosen all three times. On the third time he complained that this was not random as he was being consistently selected. Most people look for patterns in data, and when they see one they automatically assume there is a reason for it (as we discussed earlier in the difference between causality and correlation). Associated with this is an assumption that randomness does not produce patterns, so should be evenly distributed. But this is wrong. Randomness is not even; it's lumpy and inconsistent and can sometimes wander off for a long time in a particular direction that looks quite deliberate before returning. A series of coin flips does not alternate between heads and tails, even though there is a 50:50 chance of each. We will see runs of heads and runs of tails of varying lengths.

Over long times and large samples, randomness can be predictable. A radioactive nucleus decays and emits radiation randomly. A single atom could decay in two seconds or sit for a million years. However, with the trillions upon trillions of atoms in a sample, the randomness cancels itself out to become very consistent. So much so that atomic decay is used as the basis for the most accurate of clocks. The number of atoms in a sample will be some multiple of Avogadro's number, which is approximately 6×10^{23}, or 6 with 23 zeroes following – hopefully, we won't have quite this many injuries. For a large sample, there is an average that can be relied upon.

For sample sizes that are more realistic in safety terms, when a random sample wanders off away from the mean there is a tendency over time for it to return. This is termed 'regression to the mean' and can trip up analysis of data.

Consider this scenario. One year, a particular stretch of road suffers a large number of accidents. In response to this, a speed camera is installed to discourage speeding. The following year, the number of accidents reduces, thus justifying the camera's presence and encouraging similar installations elsewhere. But did the speed camera have any actual effect, or was the original increase a statistical aberration due to randomness, with the subsequent reduction simply regression to the mean? It may be that the camera was effective, but was the causation properly investigated? It may be that we never know until the next random increase. However, this generally won't be viewed as evidence that the camera was not the correct solution previously. Rather, safety systems tend to be additive and a new system will be developed in addition to the camera. Over time we can end up with multiple systems that actually cause confusion, require increased maintenance and management and generally get in the way.

In a similar way, we tend to see responses to what may be random variations in injury rates. A new system is imposed. Normal service is resumed (regression to the mean). After a while a random increase reappears and a second new system is born. System upon system upon system compounds (because we're not going to remove them to improve safety) until systemic inertia makes the whole safety process laborious and bureaucratic and the workforce is once again disengaged by the whole approach (see Chapter 14).

Impact of randomness

The impact of randomness is generally underappreciated. Consider a worker loading sheet metal and one of the sheets has a sharp section along one edge. A number of things may happen.

- He moves it without touching the sharp section – no issues.
- Introduce a random difference: he positions his hand differently. He is wearing appropriate gloves. The metal cuts the glove, but the hand is protected. If we have a good reporting culture we will get a near-miss report.
- Introduce another random element: most workers are right-handed, but this one is left-handed. This means that the dynamics of the loading operation are different and the weight is not quite as well supported. The hand near the sharp edge is carrying a little more weight and this time the steel cuts the glove and the hand, requiring first aid. We now have a first aid injury.
- Introduce another random element: last night was colder than previous nights and there was condensation on the metal. On visible surfaces this had dried, but underneath this sheet was some unnoticed condensation. Grip wasn't as good and the sheet slipped a little further. This time the cut was deeper and needed stitches. We now have a medical treatment case.
- Introduce one final random element: our worker was young and new to the workplace, inexperienced in handling the sheets. When the sheet slipped he was less able to control it than someone else may have been and the cut went deeper. This time it was sufficient to slice through a tendon, resulting in surgery, time off work and an injury reportable under local legislation.

This is obviously fairly contrived, but every day, hundreds of random factors combine to produce a set of results. The difference between no injury and a significant one can be a matter of millimetres or milliseconds, and such tiny margins can be hugely impacted by randomness. Whether a dropped object hits somebody, or a cut goes deeply, or a hammer fractures a finger, or a misaligned gasket leaks causing an explosion can be changed by whether it is raining or not, the time of day, location, wind direction, timing of walking past the event, worker height or weight, left-handed or right-handed and on and on.

It is extremely difficult to get people to acknowledge the role of randomness in safety. There seems little that we can do to manage it, so we would rather pretend it isn't there. In a world of unpredictability and lack of control, it is difficult to justify actions to take or investments to make, difficult to justify our existence as the safety person. While we may not be able to control all of these random factors, we can make our systems and processes tolerant of error and randomness. Although we can't predict what may happen, we can develop approaches that make the unexpected less likely to have a significant negative impact.

So, this is not to abrogate responsibility because we have no control over incidents, but it is important to reflect that tiny changes in complex systems can have huge impacts and we must be mindful of this when working with our data and statistics. This is a basic premise of chaos theory, known as the butterfly effect. The tiny changes made in air movement by a butterfly flapping its wings can eventually escalate into a hurricane on the other side of the world. Dekker (2011) discusses how multiple minor changes, none of which are significant or even of concern in their own right, can combine into major failures.

Action should not be taken based on results of statistics unless we are confident that they are well understood, statistically significant with clear causal relationships between the event and the action. This, of course, is much easier said than done. Few management teams will accept randomness as a reason not to act following an injury. To avoid this difficulty, as well as being smarter about statistics, we also need to establish more effective metrics. We will discuss this more in the next chapter.

Key points

From this...	... to this
• Raw data are presented in safety reports.	• Data are supported by well-founded analysis – what does this actually mean?
• Action is taken on the basis of headline information.	• Headline statistics are used as a starting point for enquiry, rather than a decision-making tool.
• Statistics are presented based on limited or poor-quality information.	• Numbers are only presented where the underlying data are of good quality and provenance.
• People compiling and reviewing information have limited knowledge of statistics.	• Information is prepared and reviewed by people with sound knowledge of statistics.

References

Blastland and Dilnot (2007) *The Tiger That Isn't*, Profile Books.
Dekker, Sidney (2011) *Drift into Failure*, CRC Press.

9 The measurement folly

'What gets measured gets done'.

This phrase, or some version of it, has been attributed to a number of different sources going all the way back to renaissance mathematician and astronomer Rheticus. Whatever its true age and provenance, over the last few decades it has been memetically embedded into pretty much every organisation in one way or another. We have myriad ways of measuring and presenting data across countless parameters to allow us to 'manage' performance.

This includes measurement of safety. Key performance indicators are the way forward. We must measure. Only then can we understand how well we're doing. How many accidents have we had? What is our total recordable injury frequency rate? How do our numbers compare against others in our industry, or in other industries? And finally, how do we use this information to properly target our improvement actions?

To improve safety performance, we must measure it so that we can track it over time.

But what should we measure and what targets should be developed as part of the measurement strategy? Many companies now have targets in place, but give little thought to what the targets are based on or, more importantly, what behaviours those targets drive.

The most commonly cited safety measure is the accident rate. These are usually categorised into 'severity' ratings. The most common ratings are first aid cases, total recordable cases and lost-time injuries. In some poorly performing industries fatalities is also a metric, but we won't go into that here because it takes little explanation and if it is a meaningful statistic (i.e. routinely not zero), there are far bigger things to worry about than injury definitions.

Definitions of ratings vary based on legal jurisdiction and from company to company, but are typically in line with the following:

- *First aid cases (FAC).* Injury requiring no more than first aid.
- *Total recordable cases (TRC).* A total figure including medical treatment injuries (requiring medical treatment by a doctor), restricted work cases (resulting in a restriction in the work the injured party is able to perform) and also encompassing lost-time injuries.

- *Lost-time injuries (LTI)*. Injury requiring time away from work – included as a component of TRC, but also separately tracked.

The principal concern with this approach is that it is overwhelmingly negative (more on this as a philosophy later). What we are measuring are *bad things*. This means that people do not want to admit to having had them. This drives under-reporting of events (where possible – some of them are difficult to hide). It takes a huge amount of work and effort in encouraging people that reporting is a good thing to develop a reporting culture where you can have confidence that everything is being identified. But, even if you genuinely want reporting (and we do, because investigation of the incident is important) and you always respond positively to individual events, there will *always* be a negative connotation when your headline graph shows that accidents = poor performance.

Now, in some respects you want your performance to be identified as bad, if it is. However, when the development of a culture is so important, as it is in safety, you really don't want such negative connotations around your principal measure.

One of the difficulties with a severity metric such as these is that the definitions are fairly arbitrary and are generally set too low for useful measures. If we are to focus on reducing serious injuries, how can we tell anything about our performance from measuring and reporting on minor injuries? We have seen from the triangular fallacy that minor injuries do not provide any supportable conclusions about significant ones in the majority of instances. We have also seen that random influences of seemingly trivial factors can make the difference between no injury and a lost-time accident. All of this is in the absence of any real understanding of what may underlie the statistics.

The way these severities are set gives us almost no information about our performance in relation to those instances we really wish to avoid. Yet there is much self-congratulation when the lines trend in the right direction. The difference between a first aid and a recordable case can be almost negligible, while the LTI category can be broad enough to encompass almost every injury in the right circumstances. In Table 9.1 are some examples of how the final categories can (and do) bear little resemblance to the significance of the incident. Reading left to right should, according to the categorisation, become increasingly severe, but if read in light of the incident rather than the consequence, it becomes somewhat less sensible. The same incident causes all the different levels of injury, but the increasing severity bears no relation to differences in the activity being undertaken. The degree of loss of control is the same in each case, but the outcomes are different. Yet we are likely to respond differently based on the outcome rather than the loss of control that prompted the accident. It is the safety equivalent of an attempted murder sentence being shorter than an actual murder sentence. What the accused attempted to do – their reasoning, their mind-set, their attitude to the law – was identical in either case, yet the sentence varies because of something that was entirely out of their control. It doesn't make any logical sense.

Table 9.1 Injury severities

Incident	Unclassified or unreported	First-aid case	Recordable case	Lost-time injury
Worker jumped rather than climbed out of a vehicle onto uneven ground.	Some initial pain, but nothing lingering.	Pain continued but minor. Ice and non-prescription pain killer used from first aid kit in the vehicle.	Pain was severe. Worker attended a local clinic and doctor gave prescription pain killer.	The following morning, the knee was swollen and tender, requiring elevation and rest. Off work for one week.
Worker slipped while cutting material in a hurry and slid the knife across the hand.	Knife didn't break the skin. No injury.	Cut hand. Suture strips applied.	Cut hand. Stitches applied.	Cut slices through a tendon. Surgery required.
High pressure blew a gauge off a gas bottle.	Missed worker's head by a few inches.	Hit worker's shoulder causing bruising. Ice applied.	Hit worker's face, chipping a tooth requiring dental treatment.	Hit worker in the eye, causing loss of sight and extended time off work.

In a similar way, injuries in the same category can be wildly different in actual impact on individuals. Broken ribs, for example, are generally not amenable to any particular treatment other than to bandage them and give them time to recover. It is not unknown to see someone with such an injury return to work immediately just with some strapping and be classified as a first aid case, although it may well be much more symptomatic of poor work practices than, say, someone taking time off work due to a twisted back sustained through some very innocuous activity.

Given such a range, how can we expect to make any sensible strategic decisions based on performance? The better organisations investigate according to potential rather than actual, but almost never is the potential rating set as a key target or metric in the business. Is a high LTI frequency rate due to several bad backs that occurred during 'normal' work, or due to several very serious injuries caused by the inappropriate activities being undertaken? It is to be hoped that if it is the latter, then some careful action is being taken irrespective of metrics, but generally our targets are based on the numerical premise that *low = good*. But is one amputation really better than three twisted ankles?

Even then, is any change due to worse or better performance, or is there some other factor involved? Have we changed the risk profile of the work we are doing? Is it a seasonal change? Is our equipment ageing, or becoming less reliable for some other reason? Rarely do the results against target come with sufficiently detailed analysis to understand these factors.

This issue becomes exacerbated with smaller organisations. Absolute numbers of injuries are generally not helpful because it gives no idea as to the rate of exposure – 100 injuries in a multinational company would be a better performance than ten on a single building site, for example. So, injuries are usually presented as a frequency rate – typically either per million hours worked, or per 200,000 hours worked (approximately equivalent to 100 workers over the course of one year).

Let's assume a business has 50 workers. This equates to approximately 100,000 hours per year. A single lost-time injury leads to a rate of 10 per million hours worked. A second injury leads to a rate of 20. This business can only ever go up in factors of ten because we don't get fractions of injuries. A typical LTI benchmark for a 'world class' company (depending on who you ask and notwithstanding the benchmarking problems we will discuss later) may be somewhere between 0.5 and 2 per million hours worked. The small company will either be at zero, where they are held up as best practice and people beat a path to their door to find out how they do it, or at 10, where they are 5–20 times higher than they would like and are being denigrated as worst in class.

But the single biggest problem with this system is the amount of time and effort that goes into category fiddling. A cut occurs. If it needs stitches, it is recordable. If suture strips can be used it is first aid. Is there any real difference between these two in terms of severity of the injury or in terms of the loss of control of the event? Absolutely not. Yet companies will expect super-visors to accompany the injured person to the doctor to make sure the minimum treatment is provided. What impression does it give to the poor worker who has been injured in the line of duty if the business is more con-cerned with categorising the injury than with treating it?

Any safety professional can reel off many stories of poor souls sent back to work dosed up on painkillers to avoid being a lost-time injury statistic. Or a bricklayer who can't do their job on the site sent into the office to shuffle paper in order to be a restricted work case rather than a lost-time incident. There are hundreds of examples – some of them quite appalling – to get around the injury classification. Yet from a genuine safety perspective, the injury has happened and what we call it has no impact on our performance. Furthermore, the effort to downgrade it may mean that it does not receive the attention it should from an investigation perspective. Safety has lost its ethical soul in the bureaucracy of scoring. Nowhere is this better illustrated than in the concern around categorising injuries taking precedence over their treatment or repetition.

Much of this is driven by the need to demonstrate safety performance to win contracts. Clients request lagging safety data as part of the tender process. Companies then do everything they can to drive the numbers down because their business success depends on it. So much for safety as the top priority.

More than any other factor in the whole panoply of safety activities, this category fiddling disengages the workforce. And rightly so. It is something that the whole safety industry – both safety professionals and general manage-ment – should be truly ashamed of. We have hurt someone and the first thought we have is how to make it look less bad than it might, even at times when doing so is to the detriment of the individual's recovery.

Safety performance goals

Many people now have safety goals tied in to their performance goals. Often, particularly at the executive level, these may be linked to bonus payments in an attempt to engender some commitment (although it is a rather perverse situation where people have to be financially incentivised to manage safety properly). The majority of targets are set around the financial year of the business in question, as year-end goals: 'By the end of this year we will have reduced our accident rate by 25 per cent.' The intention is to align with the rest of the metrics that are reported on at the same time.

What happens if there is a (possibly random) spate of incidents early in the year? There can come a point at which it is not possible to reach the goal, given the way that it has been set. What level of focus and effort is then going to be applied to safety for the remainder of the year?

My all-time favourite injury (if there can be such a thing) happened to an office worker within a group who had a zero target for the year (due to their previous year's zero performance). The poor individual ruined the target, which included a monetary bonus, by tearing a rotator cuff while performing CPR on a dummy during a first aid course.

Goal setting and targets are very difficult areas to get right. People can respond in unexpected ways and there can be many detrimental unintended consequences of poorly set goals. The potential scenarios need to be thought through very carefully. One company had a fatality gate on their company-side bonus scheme. If there was any work-related death, then the gate remained closed and nobody would receive a bonus. This is an admirable attempt to place a high priority on safety. But what would happen in the event of such a fatality? The bonus scheme would not pay out. This would result in several million dollars extra in the company coffers instead of with the employees.

What would the response from the workforce be (not to mention the public) when it became apparent that the shareholders had gained substantial extra profits due to that death? The only real solution to that problem once the gate had been established (apart from keeping everyone alive, of course) would have been to have donated the erstwhile bonus pot to charity, including the family of the victim of the accident. Thankfully (at the time of writing), this question has not had to be asked. I did raise it with the CEO, but nothing has changed.

Measurement has a strange power in an organisation. Behaviours and performance change when they are measured. This is, of course, the intent of the measurement process. Having identified the desirable performance, we set about measuring it. Over time, this provides the focus for people to increase (or decrease) whatever is being measured. You get what you look for. This very clear cause and effect requires careful consideration of what is being measured. Measure quantity, you'll get more. Measure schedule, you'll get things more quickly. Measure cost, you'll get things more cheaply. But at what other cost? There used to be a sign (in the days before internet memes) that people would print out and stick on office walls – *quality, cost or schedule. Pick two.*

One company measured the number of safety conversations as a metric. The numbers were great, but the quality suffered. It became really obvious when the cards recording the output of the conversation started being photocopied and last week's ones submitted again. It turned out on further investigation that staff were sitting in the office and filling out a card to meet their quota without having any conversations. Measure numbers – get numbers. But at what quality?

Many purchasing departments have performance measured on cost. There are numerous tales of successfully identifying a cheaper provider and meeting cost-reduction targets, only for the quality of the replacement part to be poorer and result in failures that increased risk to the workers and also increased the operating costs due to breakdowns and repairs. In one instance, what was worse was that the purchasing department measured their savings by aggregating the dollar saving over the number of units, so the increased purchasing volume due to replacements ordered for the failed parts made their cost saving look even more impressive! Measure cost – get cheap. But with what performance?

Projects are often delivered to tight schedules. The Project Manager is frequently paid a bonus tied to delivery. As a result, commissioning is often truncated. Proper checks are not made. Final tests and improvements are not completed. The project is delivered, but the operational performance is sub-par, with significant reliability issues that introduce new hazards, more stress and more cost that need to be borne by the operational team. Measure schedule – get speed. But with what long-term implications?

Another difficulty is that targets are often set via a benchmarking process that looks at other companies and compares performance. The hardest aspect of benchmarking is to compare like-for-like businesses so that the conclusion on performance is supportable. There are so many problems with the data being compared that it gives almost no valid information. Table 9.2 lists just a few of them.

By all means, carry out some benchmarking to get a high-level feel for whether you are in the right ball park, but do not make key decisions based on comparison of specific numbers. Better to build a relationship with a small number of other businesses to share ideas and lessons, so that the implications are properly understood.

Table 9.2 Benchmarking difficulties

Data set size	Typically we measure injury rates. These are, thankfully, generally quite small data sets and therefore subject to significant fluctuations. It is a general statistical premise that small data sets are untrustworthy.
Data validity	When we look at data of other companies, we cannot judge the reliability of their data. What are their reporting processes like? Do they capture all their incidents? Do they fiddle categories? Do they hide accidents because of internal targets or customer expectations?
Data consistency	Are our definitions the same as theirs? Is their version of 'lost time' any time away, a full shift only, availability for the next day, availability for the next rostered shift? Do they include contractors in their metrics or just staff?
Risk equivalence	How equivalent are their activities to ours? Even within the same industry, how similar are the risks their workers are exposed to? Is their equipment newer and better designed than ours, or vice versa? How experienced are their workers? What is the working environment like – an urban electricity network company, for example, has a very different risk profile to a rural provider, even when they notionally do the same work.
Local conditions	Accidents can be affected by innumerable local conditions that are not reflected in the headline data – weather, business performance, seasons, adjacent operations and so on.

And please, please, please *stop reporting and benchmarking lost-time injuries.* Unfortunately, this seems to be the only metric that all companies tend to use and so removing it from scorecards is met with lots of executive angst. But there is no single definition and I have seen at least four different versions of what lost time means, so comparison is often invalid anyway. And more than anything else it drives appalling behaviour in the category-fiddling scenario described above. Endless hours are wasted after events have happened trying to stop them becoming lost-time events. The accident has already happened, our loss of control has occurred. We should be spending that time understanding why and improving, not adding insult to injury (literally) by worrying more about statistics than wellbeing. Lost time is intended to be a proxy for severity. But the time taken off work following an accident is typically more related to the recovery time of the individual having the accident (due to age, general fitness, etc.); the medical professional that decides whether time off work is necessary or not or the receptiveness of the particular injury type to treatment. It actually tells us very little about the degree of loss of control suffered.

Safety professionals agree, almost without exception, that it is a terrible metric, but nobody seems to be prepared to drop it. I suggested this to one company during the development of their new executive and board safety dashboard. The response was that in their benchmarking enquiries, this was the only metric that everyone measured. So, they were only looking at it because everyone else did. So far, that is the only reason I've been able to ascertain. Everyone thinks it's pointless, but everyone else does it, so we'll carry on. Someone needs to show some leadership and break this cycle. The sooner we do, the better.

What to measure?

Safety is also unusual in that its main metric is trying to demonstrate something by its absence. We attempt to show safety by measuring accidents that occur due to the lack of safety. This is the equivalent of trying to measure revenue by counting how much money customers spend on purchases other than our product. Or a government measuring population growth by how many people decided to stay away. There are inferences that can be made from these non-statistics, but it is tortuous and not intuitive. What we're measuring is not safety but *unsafety*. And, even then, we're not measuring it very effectively because it is easy (and common) to act unsafely without any injury occurring. At the very best, we can say that we're not terrible. But we can't measure how good we are. It simply makes no sense to try to measure something by the absence of its opposite.

It's a little like measuring the success of a golfer by how many putts are missed. It tells us nothing about approach play; about whether the putts were difficult or easy to make or how many were attempted in total. A poor player who plays only one hole with one lucky shot on the green will seem better than the one who misses only a single putt in an entire 18-hole round.

Have you ever used a knife instead of a screwdriver at home when changing a plug on an electrical appliance in the kitchen, or maybe even just tightening a pan handle? Or used one to retrieve something stuck in the toaster? Or broken the speed limit while driving? (Go on, admit it; no-one else is listening.) None of these are necessarily 'safe' and yet we have all done them without injury. So what exactly is our injury performance telling us about how safe our activities are?

Not only is it illogical to measure in this fashion, it is also demoralising for the workforce.

It is widely understood (although not always practised) that motivation comes from positive sources. It is frustrating and damaging for a team to work hard to deliver only for management to focus on the small numbers of times when something went wrong. There are thousands of management books, self-help books, motivational speakers and good parents out there who know this and preach positive reinforcement as the method to improve performance. Yet this negativity is not only present in almost all companies around safety measurements, it is institutionalised in company reporting and actively encouraged.

Organisational psychologists will tell you that you get what you focus on. The messages we send should be focused on solutions, not problems. The brain thinks in pictures and moves towards whatever image it is frequently sent. Thus negative reinforcement generates a negative image and leads us towards it. Positive reinforcement does the same in the opposite direction. This is now recognised in many areas – fire signs now tell people to 'stay calm' and 'walk', rather than 'don't panic' and 'don't run'. This prompts an image that helps in the event of a fire. What message is our collective subconscious hearing when it is bombarded consistently with stories of the number of accidents, failures and injuries we're recording?

It is widely known within the safety profession that 'leading' indicators are preferred to the lagging ones we have discussed. These look for activities that lead to safe outcomes and monitor them – leading the event. These have been used and discussed for at least 20 years (Sefton 1997), yet even now it is a struggle to get executives and boards to accept the use of leading indicators as a principal metric for safety.

Typical leading indicators include safety training, safety briefings, safety conversations, completion of audits and inspections, speed of response to hazards being raised and near-miss reporting (although there is some argument as to whether this is a lead or lag indicator – just measure it and don't argue over semantics). There are many others, and they can be chosen to reflect the activities and risks of the business in question. *Step Change in Safety*, an initiative begun by a group of oil and gas companies operating in the North Sea, has produced a helpful guide. The same considerations around unintended consequences apply as discussed above, but focusing on what is safe is both more logical and more effective.

Leading indicators need to be carefully considered and made appropriate to your business. When developing a set of indicators, there are two key considerations:

1 Is this a valid data set that is giving me useful information?
2 What am I going to do with the information once I have it?

Too much time is spent measuring information that either doesn't tell us anything, or that we don't (or can't) act on to make any clear improvements.

Hollnagel (2014) has taken this positive focus a step further in his development of *Safety I and Safety II*. This suggests a total focus on the positives of the good performance we see every day and in 99+ per cent of activities, understanding and isolating the components of what we do well, so they can be learned and promulgated. Or carrying out formal investigations (rather than just a lessons learned review) into particularly good performance.

Businesses must move themselves away from negative and illogical reporting metrics if they are to provide a supportive environment for safety improvement.

Reporting it

Having determined what leading indicators you are going to measure and monitor, you will need to report the output to management. In determining what that reporting should look like, you should bear in mind some of the issues highlighted in Chapter 8.

The key thing to note here is that the safety team (once we've made sure they are armed with the required statistical capability) should interpret the data and present that interpretation alongside it. This does not have to be complex, just a high-level analysis of the data and, where necessary, a note describing the action to be taken in response.

For our misleading rolling average graph in the last chapter, for example, we might simply add a statement that says: 'This month's increase is due to a statistical outlier a year ago early in the data set. The underlying trend remains positive. No action is required.'

This short analysis first prevents the executive from jumping to conclusions or to knee-jerk reactions, and second, provides them with the confidence that performance is being well-managed.

We should not be afraid to provide much more extensive executive and board reporting on safety. As information passes up the organisation it tends to be summarised more at each level. By the time it reaches the board, we typically have little more than a smiley or a sad face. For most businesses, the annual report will contain a statement about the overall injury rates across two or maybe three categories, and that is it. I recently reviewed an annual report of a major multinational oil company. The safety report comprised three pages of information. Yet there were six pages simply telling us who the company directors were. Compare this with the financial report. While there are a couple of headline categories – obviously revenue and profit – it is widely recognised that these can be very misleading regarding the actual financial health of the business and so they are accompanied by a myriad of other supporting statistics to allow for a more realistic understanding. These include profit as a percentage of revenue; underlying profit after removing exceptional items; debt to earnings ratios; cash on hand; current work in progress; days sales outstanding; asset to liability ratios; assets per share and so on.

This wide range of statistics begins to paint a proper picture of company financial performance. A similar level of reporting rigour is required to properly understand safety performance.

Key points

From this…	…to this
• Basic lagging statistics are the key reporting metrics.	• Reported metrics include relevant leading indicators and incidents focused on potential.
• Accident response is driven by the category of the accident.	• Accident response is driven by the best interests of the injured party and the level of control lost in the event.
• Lost-time injury is the key metric.	• Lost-time injury is no longer recorded or reported.
• Individual targets are set with little thought as to the unintended consequences.	• Targets are set to drive specific activities and behaviours, not linked to outputs.

References

Hollnagel, Erik (2014) *Safety-I and Safety-II: The Past and Future of Safety Management*, Ashgate.

Sefton, A. (1997) Leading Indicators – Safety Measurement in the Future. Opening address at the International Association of Drilling Contractors Seminar, Aberdeen.

Step Change in Safety (n.d.) Leading Performance Indicators, a Guide for Effective Use.

10 The communication gap

Much is made of the role of leadership in driving safety performance – good and bad. And quite rightly so. Positive, thoughtful and intelligent intervention, support and commitment from the leadership makes a huge difference in safety performance. As it does in every facet of running a business.

Many companies spend a considerable amount of time in conversation between their management and their safety teams. Presenting metrics; doing safety walkarounds; holding safety committees and reporting on incidents and their investigations. Yet virtually all of this communication is of poor quality. There is little critical examination of information. Little genuine challenge and constructive debate. In short, we're all spending lots of time getting nowhere fast, all the while thinking that we're achieving something.

One of the most commonly repeated complaints from the safety professional is a lack of support from management. They don't give enough airtime; they don't provide enough support; they never commit sufficient funds and resources; they see safety as a cost, not an investment.

The counter to this argument is the most common complaint from management about the safety professional – they don't understand the business and how it has to manage its priorities; they see safety as the only thing that matters; they never understand the working impact of what they're asking for; they never explain the benefits of their requirement; their focus is too narrow.

The solution put forward most frequently to this problem, presented by management and safety professionals alike, is the need to explain safety in broader business terms – perhaps by attaching costs to it, quantifying the benefit to be gained, showing a return on investment. Dollar values are easy to understand. In a nutshell, safety professionals need to learn to speak Boardroom.

To an extent this is true, but there must be a far greater emphasis on leadership to speak Safety. If you are in an executive leadership role, such as a board director or CEO, it is simply unacceptable for you to be a safety illiterate. If you genuinely believe that safety is a high priority for your business, how do you reconcile that with the fact that you don't understand it? Would you accept from someone in your position that it was okay to be unable to understand a balance sheet? Not just the numbers or headlines presented, which most people can cope with, but with sufficient insight to be able to ask penetrating questions about the health of the business based on what you're seeing.

If this is the case for finance, why not safety? I'm sure most managers know what LTI stands for and perhaps what their main investigation tool is, but how many can engage confidently in a discussion around the less obvious but more important factors such as risk perception, human factors, motivational and behavioural theory?

Communication is a two-way street. Both management and the safety profession are very poor at communicating about safety, but in very different ways.

Safety professional

When interviewing for safety roles, the closest thing I have to a go/no-go question is: 'What is the single most important skill a safety professional needs?' The correct answer is communication skill. Not everyone will agree with this as the most important thing, of course, but everyone agrees that good communication skills are a key component in the safety professional's arsenal.

There are three key components to a safety manager's role in an organisation, as shown in Figure 10.1. First, they act as a subject matter expert. When the business has a difficult safety problem to deal with, they can come to the safety manager for expert advice – what is the chemical with the least harmful characteristics, for example, or how should we manage this asbestos we've

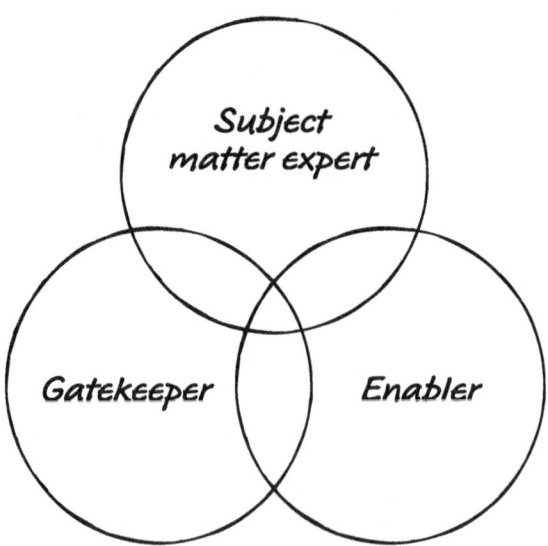

Figure 10.1 Roles of the safety professional.

just inadvertently come across? Second, they are an enabler. They provide suitable tools, processes and approaches to enable the business to manage its work safely. The majority of work should be able to be conducted within the core skills of the operations teams and not require the intervention of the subject matter expert, but this requires those tools to be in place. Finally, they act as a gatekeeper. In matters of safety, they are the person that stops the business from exposing its workforce to unreasonable levels of risk (in theory, this should be unnecessary in a well-managed organisation, but the reality is that a degree of independent oversight will always be required).

This is an extremely delicate balance to get right and is difficult to do well. Yesterday they were helping you and today they're stopping you. It requires a high degree of trust in the safety manager for others to accept their other roles in combination with that of gatekeeper. This can only be established on a bedrock of strong communication. It is a vital requirement. But we still continue to see so many instances of safety professionals with poor performance in this area – whether failing to get management on board, or spending all their time in the office and not interacting with the workforce.

There are a number of reasons for this. Many safety professionals come from working on the tools in their past, while others are career safety professionals that have limited experience in any other field. This is a gross generalisation, of course, but neither circumstance lends itself to naturally broad communication skills. Shop floor converts tend to have difficulty in dealing with higher levels of management due to a lack of exposure to them, while career types – like many other specialists – can often be too engrossed in their own area to adequately understand the more wide-ranging implications of their discussions.

Furthermore, typical safety training courses cover safety legislation, Heinrich's triangle (unfortunately), investigation techniques, hierarchy of hazard controls and so on, with only a small introduction to communication skills, if any. Ideally, significant specialist training in communication should be undertaken, in conjunction with practical experience, to develop the skill levels required. However, most businesses tend to be reluctant to fund such training for their safety team. This issue is not restricted to safety. Ask most people what skills were demonstrated by the best project managers they have ever worked with and their answer will typically include communication, person-management, motivation and other similar attributes. But go on a project management course and what do you learn? Gantt charts, scheduling, contingency management, Monte Carlo analysis of project risks. These things matter, but without the accompanying 'soft' skills they cannot be properly deployed.

The higher up the management ladder, the more important these skills become. Senior safety professionals need to be well-rounded, with broad experience of different industries and preferably different functions as well. There is a need for industry- (and even organisation-) specific specialists, too – in an underground mine for example, I would want to know the safety manager knew *a lot* about mining – but the current gap appears to be in the well-rounded part of the equation. To gain a seat at the executive table and to be able to influence effectively, individuals need to offer more than just their technical background. They need to understand the business as a whole and contribute usefully to those business-wide items on the corporate agenda.

There is a risk of this becoming worse as more and more organisations and, in some cases, governments, are setting out minimum levels of training and experience for safety professionals. This makes some sense, but there is a danger of this becoming a cumulative requirement and demanding more senior people having increasingly higher levels of experience and qualification of *a specific health and safety nature*. In reality, the higher up in an organisation you are, the less functionally specific is the nature of the job. Senior managers are, by and large, generalists. Overemphasis on direct experience and specific postgraduate-level qualifications will tend to discourage people transitioning into safety later in their careers when they have already reached a senior organisational level and actually may have some very valuable insights.

Links to some of the other areas in this book also provide some of the roots of the communication gap from the safety side. It is difficult to gain buy-in from management when some of the data being presented falls well short of any meaningful analysis of the situation. Telling people that safety should always be the number one priority causes great difficulty when the board has to balance that against other competing requirements (despite what the official line on safety priority may be). Returning again and again asking for support when previous initiatives cannot be demonstrated to have worked leads to a significant credibility shortfall that takes a long time and a lot of effort to recover from. When leaders undertake site visits to hear front-line workers complain about the process overload due to safety, it becomes difficult to support.

In addition, the majority of (good) senior leaders are accomplished critical and strategic thinkers, even if they don't have specific safety knowledge. Conversely, many safety professionals are guilty of a lack of strategic thinking. This immediately puts them on a course that is at odds with senior leadership, who are required to have longer-term horizons. Most safety strategies look out to a year ahead at most. I once attended a meeting with HSE managers from the same company across Australia and New Zealand who were presenting their strategies. Not one of them looked beyond a single year ahead. This was an organisation which was on the brink of some significant changes in the type of business in which it was engaging. Changes that would result in an increase in the exposure of their workforce. This level of disconnect with the general thinking of the executive can only lead to communication failures.

Safety strategy should cover a 3–5-year horizon and should be closely linked to the overall business strategy. Whether the organisation is expanding, contracting, diversifying, acquiring or divesting plays a large role in the potential future risk profile. It is no good acquiring a new business in a different industry as part of a diversification strategy and waiting for accidents to work out how we should change our safety approach.

Again, this lack of strategic vision is not restricted to the safety profession, but in an industry where under-performance is deeply rooted in culture and behaviour that is well known to take significant lag times for change to be effective, it is almost inconceivable to have a strategy that is not at least three years in duration, let alone no real strategy at all.

Management

Safety professionals do need to learn to speak Boardroom, but they also need to be met half way.

As noted at the beginning of the book, lack of support from management is not generally a lack of caring. Managers do not have emotion removal operations on reaching a certain level in the organisation. What management lacks is a full understanding of the matter and a clear idea of the risks being faced. One member of a leadership team that I worked with transformed in a short time from being the least engaged of the team in safety to one of the most. The difference was visiting two workers in hospital in the space of three months who had been seriously injured on projects in his area. I'm not suggesting we should hurt people to make management aware of the potential, but managers do need to spend more time in the field. It is easy when working at Head Office, spending all day in meeting rooms, to forget the very real risks that the workforce is exposed to.

When people gain a better understanding, they become more engaged and will develop their own knowledge further.

A significant part of the management problem is that they do not always understand the impact of their communication – not in what they say, but in what they do.

> What you are doing speaks so loudly that I cannot hear what you say.
> (Ralph Waldo Emerson)

Managers often do not understand enough about safety to be able to take a sincere interest. There are few probing and insightful questions. Few challenges to data or investigation findings. Little detailed follow up. Compare this to their areas of expertise. When they are presented with financial data, for example, they consider it carefully and ask an abundance of questions. It may not be their intent, but what they are doing is demonstrating much less interest in safety than other areas. The unspoken implication is that safety is less important and of less interest, so the business will respond in kind.

Management are often trying to do the right thing and will say what they believe to send the right message. But often they do not support the words with actions – walk the talk – and this incongruence undermines the best intentions and disengages the workforce. It is important to be mindful of the message that is being heard, not what is being said, as the two can be very different (see Chapter 18).

People in senior roles often feel that they need to have the answers, so when they have little knowledge, they won't explore in detail. Asking questions shows a degree of vulnerability that many people are afraid of, but which is actually very powerful in leadership terms. The irony is that the best way to show interest is to ask questions. This is much more impactful than telling, and it also increases understanding.

Organisational structure can also have a significant bearing both on the communication channels that are available and also on how safety is perceived. Some people suggest that the Safety Manager should always have a direct reporting line to the CEO. I don't necessarily subscribe to that – it depends on the size and complexity of the business and the level of risk it faces. But someone on the executive team should have direct responsibility for safety and that person should have a good level of knowledge on the subject.

For some reason, safety is often included in the HR portfolio. I can only assume that this is due to some vague 'it's all about people and their wellbeing' connection, but I can genuinely see no real reason behind it and have yet to have anyone give me a compelling answer to the question. Take a look at the biographies of a few leadership teams on company websites. By and large the HR Manager/Director is the youngest there, with the least experience in industry and almost none of that experience will ever come from any role that has significant exposure to workplace hazards. This may possibly be acceptable in a services organisation where all staff are office based. But in any kind of industrial company with genuine high-risk activities, the person with executive responsibility for safety must have a good knowledge of the hazards faced. This would be most likely to come from operations, engineering, asset management or maintenance areas. While there is an argument for a degree of independence between the safety reporting line and the operational one, this is generally in relatively immature organisations and comes a distant second to the requirement for expertise in the role. On a wishlist for responsibility for safety it would be a close run thing between the HR Manager and the CFO for last place.

This is another good question for someone who exhibits the Priority Confusion: 'If safety is your number one priority, why is there no-one on your executive team with specialist safety knowledge?'

Note that this does not preclude some capable individuals from HR doing a good job in the role (and they are out there). But even then they are always disadvantaged due to their lack of intimate knowledge. Furthermore, it undermines the role of the safety professional, who must always be subordinate to a different department, with an extra layer of management between them and the executive team, than if they were an autonomous function. It tends not to be in the specific safety discussions where this lack of knowledge gets found out – the HR Manager can prepare themselves well for those discussions, or bring along their subject matter expert for the discussion. Rather, it is in recognising the safety implications that may be there in other decisions that are made by the executive – in understanding the safety implications of a proposed course of action.

Management need to do a much better job of understanding the intricacies of safety and, more than anything else that they do, ensure the message that they are sending will be consistent with the future actions they may take. Any inconsistency will derail the best intentions. This is not to say that messages should be untruthful or subject to 'spin' to sound different to reality. It is simply to ensure that the message is delivered in a way that is authentic, congruent and liable to be well-received.

There is also far too little communication and not enough transparency in what information is provided (again this could be levelled at the whole corporate world, not just safety). Communication needs to be much broader than simply about safety as there is a requirement for the business to understand the context in which decisions are made at senior levels. There seems to be a reluctance to give people general information, with most briefs/emails targeting only those staff that appear to be directly affected by any particular issue. The outcome is a general bewilderment at decisions that are made that appear to have little or no grounding in anything remotely sensible and with apparently no forethought as to the operational impacts. Often this is because people do not know the context and so cannot understand it. A little extra communication and openness would help to alleviate this hugely.

I had a French teacher who taught us that if we sounded to ourselves that our accent was ridiculously over the top and threatened to become an insulting caricature, then we were probably just about getting it right for the listener. This has parallels with communication in business. If you think you are going way over the top and doing too much, it's probably just about okay for your audience. If you think you're doing a reasonable job, it's not enough.

It has to be recognised that corporate communication is a no-win situation. One person's desire for more interaction, who pretty much wants the CEO to fill them in on all the details personally, is in direct conflict with one of their colleagues who desires to be left alone. Before undertaking any employee survey, it is almost guaranteed that poor communication will be one of the issues raised.

Key points

From this ...

- Managers know little about safety. Safety managers know little about business.

- Management say what they think they need to say but don't back it up with actions.
- Safety strategies are really just annual plans.

- Management occasionally talks about safety when prompted by events or activities.

... to this

- General managers and safety managers have a well-rounded knowledge of safety and its role in the broader context of the business.

- Management are aware of the non-verbal messages they send and how they are perceived.
- Safety strategies cover 3–5 years and align to the broader business strategy.
- Management errs on the side of over-communication.

11 The zero paradox

'Our goal is a zero-harm workplace'.

Fine words and a marvellous sentiment. Wouldn't it be great if nobody ever got hurt?

The zero-harm movement has gained significant traction over the last few years. At face value, it seems to be the right approach. It's hard-line – we're not prepared to accept anything other than the best. It shows how caring we are – we don't want anyone to get hurt on our watch. It has an unpleasant alternative – if our target is not zero, how many people are we planning on hurting?

As we have seen throughout this book, safety is a complex beast. And we should be very wary of anything that seems very simple and absolute. How can we be absolute about safety in a world where there are substantial aspects over which we have little or no control?

What is zero harm?

The statement that often accompanies zero harm is that it is unacceptable for anyone to be harmed in the workplace. But what does that mean and where does it come from? Does it mean any harm at all? What about the old cliché about paper cuts in offices? Is that unacceptable? What if I walk into a desk and bruise my leg? Is that unacceptable?

It doesn't seem that there is very much concern when either of these happen (and justifiably so). Why is that if zero harm is the goal? It would appear that there are two principal reasons. First, that the degree of harm is so low as to be negligible and not really categorised as harm at all. Second, that it is virtually (if not completely) impossible to manage our environment to the extent that complete elimination of all such injuries is credible. We do away with paper to prevent cuts and suffer from carpal tunnel syndrome from using the computer instead.

All organisations accept these risks, so what is zero harm? It appears to be the absence of all injuries, but for practical purposes it is subject to some arbitrary cut-off point.

But where is the cut-off point at which an injury becomes unacceptable? Referring back to our categories in Chapter 9, is it a first aid case, or is it a medical treatment case? Who makes the decision anyway? One pro-zero argument is that anything other than zero is somehow morally wrong. Has anybody been to the workforce and asked them what is acceptable to them in the course of their work? Did anyone ever embark on a career as a carpenter with thoughts that it was unacceptable to cut their finger, or hit it with a hammer? If so, we would have no carpenters.

Acceptability

Who has asked the people taking the risk what risk they are prepared to accept?

Consider New Zealand, the home of the All Blacks, the national rugby union team that is one of the most successful teams in history in any sport. Rugby plays a huge role in the New Zealand national psyche.

In 2011, the All Blacks played in the final of the Rugby World Cup. Consistently the world's number one team, they had somehow failed to win the trophy for 24 years. This time it was on their home ground. The final was against France, a team with a history of upsetting the odds in these situations, defeating the All Blacks in both the 1999 and 2007 competitions. Richie McCaw, the New Zealand captain and arguably the greatest player ever, led the team and the country through anguish and trauma in a game that was far closer and more uncomfortable than it ought to be. But they won and the nation rejoiced.

It later emerged that McCaw played the game with a fractured foot, playing through the pain to bring the trophy home. The response to this across the fans and the media was unanimous: McCaw the brave, the warrior, the hero.

It is a well-known narrative in sport with numerous iconic photographs of players soaked in blood giving their all for the cause. But how would this play out in a workplace setting? McCaw is a professional sportsman – the pitch is his workplace and the New Zealand Rugby Union his employer. Was the regulator notified of the broken foot as required under New Zealand law at that time (or possibly required – subject to interpretation)? Was the Union acting illegally in sanctioning his involvement in the knowledge that there was a high risk of worsening his injury (if they knew at the time)?

I'm not for one moment advocating greater involvement of lawyers and regulators in professional sport (please, no). But how can one workplace operate like this when another (in fact most others) puts significant time and effort into preventing any injury of any type or severity and campaigning for zero harm? (Of course, many seem happy to bring someone back *after* an injury in the same way – those cynics among us may suggest that this is not necessarily for the right reasons, but that is a different discussion. See Chapter 9.)

Different workplaces have different risk profiles and so require different interpretations, which is not in itself an issue. On the surface, those profiles can explain the difference in approach. There is also a degree of recognition (and acceptance) by people in high-risk occupations that they are exposing themselves to a greater extent. No-one expects a police officer or a soldier to have the same approach to risk as a bank clerk.

But those people who revere McCaw and similar heroic figures in sport, in movies or in the military, are the same people who are at work in our workplaces every day. In many cases, they also play the sport and expose themselves to similar risks in a non-work setting, albeit at a somewhat lower intensity than elite players. They spend 80 minutes of every Saturday afternoon kicking lumps out of each other on the rugby pitch, proudly carrying scars of battle, walking off with shirts drenched in blood sometimes, but not always, their own. All in the name of enjoyment.

Although their risk profile at work may be low, their risk appetite is shaped by factors outside of work – cultural, personal and societal expectations that are much more strongly felt than their occupational equivalent. For them, an injury may be no big deal, maybe even a badge of honour.

This leads to a discrepancy between worker expectations and company expectations. As we have noted, the company approach is usually captured in a statement such as: 'No task is so important that it is worth an injury.' The workers' view, however, is closer to: 'Sure it is. It just depends on what the task is and how big the injury is.' Breaking a foot to win a world cup is an acceptable outcome to most sports fans. In fact, particularly in a contact sport, someone not prepared to accept that injury would be derided.

Everybody has a different perspective of what is an acceptable risk to take. Most people would accept a larger risk to save a child than they would to save a dog. But that level of risk varies according to the situation – is it their child, or their dog, or someone else's? Or according to their background – do they come from a family with a history of family pets, or were they once attacked by a neighbour's dog?

When there is a mismatch in views like this, the most powerful person usually prevails. In this case, the company. So the worker complies with the controls for 'trivial' risks, but does so unwillingly, begrudgingly and with a steadily growing disdain and resentment for all things safety. When higher risks later become apparent, the worker is less likely to fully engage in the risk management process.

What response do you think you would get if you offered them the choice between a cut finger once a year and a mountain of paperwork every day to assess their safety exposure? They would choose the injury and they would do it with a degree of emphasis that would involve colourful language not for publication.

That is not to suggest that we should be blasé about safety or that any injury is okay. It is also reasonable that an employer paying wages can expect the worker to conform to the company view of risk, but when we need the engagement of the workforce, they do not wish to be mollycoddled and patronised. They want to be protected from significant risks, not from trivia. There is a general societal backlash occurring about the overprotective nature of our modern parenting and government. Nobody is allowed to take risks, so they never learn from them and they never experience the positives around risk taking. This is why there has been such an explosion of participation in extreme sports. People have nowhere else to go to get some excitement in their lives. And the lack of minor risks is driving the search for ever more extreme ones.

What happens if we have an accident that requires no medical treatment or restricted work? It is quite conceivable for a serious event to fall into this category as we have previously discussed in the case of broken ribs. The same is true of broken fingers, for example. In many organisations this could happen while still celebrating zero harm. This situation would reek of hypocrisy to the workforce.

I have lost count of the number of times I have heard from workers that the important things are ignored while we obsess over trivial details. The workers understand their environment. They know what risks count. Zero harm makes no sense to them.

Similarly, something major can happen but without injury. When the Buncefield oil storage facility in the UK exploded in 2005, nobody was hurt. They successfully achieved zero harm through sheer dumb luck that the explosion occurred outside of working hours when nobody was present. As an extreme version, this beautifully illustrates the point that zero harm can be achieved even when operating very unsafely. As a goal, target or vision, it does not achieve its intent.

How realistic is it?

Is it achievable? Zero-harm proponents would have you believe that if you can go one day without injury, why not five? Why not a week? A year? And so on to eternity. And when we see celebrations for ten million hours worked without an injury it seems tempting to believe that it could be achievable. There is a certain logic to the argument. But the stark reality is that over a long enough time period, or across a large enough population, there are simply too many uncertainties and too many uncontrollable factors. At some point, an injury will occur. It is inevitable. And then we will, despite all our best efforts, have failed. And it is this inevitability that makes people cynical, seeing zero harm as a slogan only, as lip service, rather than as something real and achievable.

This failure is, in some respects, a foregone conclusion. It may not happen today, or tomorrow, but it will happen. Reality will not let it be otherwise. Take an employee walking across the company car park between buildings. A sudden, unexpected gust of wind blows a foreign body into their eye. Medical treatment is required to remove it. We have lost our zero result due to something over which we had absolutely no realistic control. Feel free to try to get budget approval (from your 'safety is our number one priority' executive) to enclose the entire car park in an overbuilding with filtered inlets to prevent this incident and see how that is received.

The fundamental basis to all goal setting is that the goal must be achievable. Why should safety be any different to every other goal we set? Can you imagine a business that said it was not prepared to countenance any activity that resulted in a loss, no matter how small? Zero losses across all new product launches as the only acceptable outcome?

Of course, the counter to this inevitability argument is that aiming for zero at least reduces the likelihood and is, therefore, better than having some other goal. This has two key flaws. First, it assumes that the goal must be replaced by some other injury number that is not zero. But there are plenty of better ways to measure safety and to provide goals than the number of injuries. An oft-repeated pro-zero position is: 'If your goal is not zero, how many people are you planning on injuring this year?' This is a puerile and over-simplistic argument. Safety cannot be dumbed down into polarised black-or-white positions. Just because something is not achievable does not mean you are endeavouring to achieve the exact opposite. I recently had the same conversation in a workshop with a zero-harm proponent who presented a similar argument. The following morning he gave me a lift and took a call on his mobile phone while driving (until I suggested he pull over). So much for the zero-harm culture in his organisation.

Imagine the following conversation:

'Is your goal zero harm? With no injury ever to happen to anybody? We have done it for six months – it must be possible to go all year.'

'No. Actually seems unreasonable to me. We can't control all risks. I'll be happy knowing we've controlled the really serious ones as best we can.'

'If your goal is not zero, how many workers are you planning to hurt? Have you told them?'

This is not too dissimilar to many discussions I have had and heard, with the questioner taking the moral and ethical high ground. Now let's put something similar in a non-safety context, say an investment fund manager.

'Presumably you are gunning for zero losses in your capital investment?'

'Well, no. Not every venture can succeed. But on balance I think we can be reasonably profitable even if there are some that don't work out this financial year.'

'What? How many are you planning to fail? Have you told your investors?'

Such a conversation would be unthinkable in financial markets. The person demanding no failures is completely unreasonable, given the complexities and vagaries of the market. Yet, swap the money for work-related injuries and all of a sudden the unreasonable position seems to turn around according to many people.

Of course, in safety there is an ethical and moral perspective that may not be present in other areas. One promotional video for road safety in Australia asked someone how many deaths were acceptable on the roads. When he suggested a number based on a reduction from current levels, the same sized group of people was introduced made up of his own family and friends. With visions of his own family as the victims and with tears in his eyes, he changed his view to zero. This emotional approach is very striking and impactful. But, does it actually make a difference? Driving is largely a skill-based activity in which people rely on experience and carry it out semi-automatically. Drivers are generally aware of the risk of crashing, but that knowledge does little to change their behaviour. Reminding them of it again would seem to have little effect. The money spent on producing and distributing the video may well have had more life-saving impact spent on physical works improving visibility at a particularly accident-prone intersection.

Being reminded that harm is bad has limited impact on how we actually carry out our work.

The second flaw is that a stated goal that is unachievable has no motivational value. The entire workforce listens to the goal, knows it is not achievable and switches off. Worse still, the sight of an executive stating their firm belief in zero harm, in spite of the clear logical fallacy, destroys any credibility they may have in relation to safety. Once again this disengages the workforce and actively acts against improving safety. Some companies prefer a message that says 'safely home every day'. While ultimately this suffers from the same logical failure in that we can't all go home safely every day forever, at least the workforce can start each day thinking it is achievable today. They would also probably consider they are going home safely even if they had a stitch or two – so long as they had all their limbs intact and nothing too important was broken. The way it is phrased leaves some room for the grey areas of life. It allows for interpretation and discussion, and, ultimately, that leads to engagement.

The other argument that is usually put forward is that zero harm is not a goal or target, so much as it is a vision or a philosophy. I have some sympathy with this position. After all, we are here to stop people from being hurt. But if you have to explain to your workforce that it's a vision not a goal (corporate speak that most of them don't want to listen to anyway) and how they need to interpret it to understand, then I would respectfully suggest you've already lost the communication battle.

The impact

People feel they are being patronised, not protected, when they hear the zero-harm message. There is at least one major international company with a stated zero-harm philosophy that has a different cut-off point for local sites than it does for regions. For a site to achieve zero harm, there must be no medical treatment cases. For the region, there must be no lost-time injuries. I found this out when I challenged their performance poster that had more zero-harm days successfully achieved for the region than for one of the sites within it. This is not possible with the same target, so I assumed there was a mistake and helpfully pointed it out. But there was no mistake. It was quite deliberate. This difference is showing that achieving zero harm for a larger population is harder than for a smaller one – a tacit admission that the inevitability argument holds true.

Because it is harder at a larger level, they didn't try harder, or recognise the flaw and change their philosophy to a more appropriate one. No, *they changed the definition to fit.* And then stuck it on the wall for everyone to see. These are exactly the kind of verbal gymnastics that companies perform to try to justify an impossible and absolutist target. And it undermines all credibility in that target.

What happens to the person who suffered the injury that was sufficient for the site but not the region? First, they have been hurt. Second, they get a hard time for ruining the site's zero record (another side-effect of zero targets is under-reporting of injuries to avoid this). Neither of which is much fun. Then they see the regional management getting a pat on the back for achieving zero harm. How does that make them feel? Will they be appreciative of the performance? Or be hugely negative about the whole approach and broadcast that message to everyone prepared to listen?

What happens if we do reach zero for a while? How do we develop and maintain continuous improvement when we have met an absolute target? We may possibly become complacent and do nothing, thus making it likely to slip backwards or, if we are a forward-thinking organisation, we probably look at reducing the number of near misses and measuring that. Which highlights the most serious failing of the zero-harm ideology. Looking in all the wrong places.

As we discussed in the triangular fallacy, there is little by way of causal links between most near misses and significant injuries. Reducing near misses and reducing minor injuries does not, in most cases, help to reduce serious injuries or fatalities. If our goal is zero injuries, there is a drive to reduce numbers. Ninety-nine per cent of all injuries are minor and will, when driven by number reduction, receive 99 per cent of the attention. While we are busy stopping people cutting themselves or hitting themselves with hammers, or investigating near misses where someone tripped on a walkway, serious injuries are lying in wait.

This becomes particularly ridiculous at and around the cut-off points between acceptable and unacceptable harm. As we have seen, if a worker cuts a finger and requires stitches it is, in many instances, a recordable injury and results in a failure to meet zero. But if we can treat the cut with suture strips instead, it is only first aid treatment, in which case, bizarrely, we meet zero even though we have had essentially the same injury. The choice of whether to stitch or not is, notionally, a medical decision but actually it's very subjective. Different doctors may make different decisions. So the outcome of our zero target in this instance bears no relation to the actual injury. It is being decided not by the severity of the injury, or the potential that it could have had under other circumstances, but by a third party's personal preference. Much effort is expended on managing not the injury, but the nature of treatment (including sending a supervisor along for treatment to make sure it's as little as possible – hugely demotivating for the employee). This is effort that could be far better spent looking at more serious risks.

Assuming an 80/20 Pareto relationship, 80 per cent of our attention should go on the 20 per cent of injuries that cause 80 per cent of the harm and trauma. While these numbers are simply for effect and not a statistical basis for discussion, the concept is correct. We get what we focus on. Ask for fewer injuries, we will get fewer injuries, but we will not be discriminating which ones we work with. In fact, the focus is such that the easy ones will get the attention. In a zero-harm world, reducing from 100 injuries to 1 is seen as becoming successful (not quite there, of course). But if the one that remains is by far the most serious, we have achieved virtually no material risk reduction at all.

The zero-harm world does not acknowledge risk or context. I was once embarrassed to receive a zero-harm award for my team of a few consultants who sat behind desks all day working on computers – exposed to potential scalding events from coffee. The reason for the embarrassment was the lack of award for other colleagues in a different department. They had executed over one million hours of field work in high-risk environments under constant delivery pressure. But a single (minor) accident had blotted their copybook. Their achievement was orders of magnitude more impressive than ours, but we got the award due to the overly simplistic, risk-agnostic approach taken. It is the equivalent of a safety award for a fire service call centre operator, while denying one to the firefighter who saved lives but received a burn in the process.

There are too many work-related fatalities, but they are nevertheless relatively infrequent events. When quantifying risks, fatality acceptability cut-offs for workers typically have a probability of fatality somewhere in the region of 1 in 10,000 per year. Even for a large workforce on a site of 1,000 people, that equates to one fatality every ten years. It is quite possible to go for a year, or two, or ten, or more with no significant injuries and assume that your performance is good, only for the law of averages to come and kick you in the teeth. The absence of injuries does not prove the presence of safety, as we mentioned earlier.

That bears repeating, though. *The absence of injuries does not prove the presence of safety.*

If you want to know if something is being done safely, you need to understand what that safety looks like and make sure it is happening. Not look back after the fact and see if everything turned out okay. This is in direct contradiction to the zero-harm position. There is almost universal acceptance that lagging indicators are less preferable than leading, because they give us limited prevention information. But zero is the ultimate lagging indicator. It is only available after the fact and it may well mask serious issues that are not arising due to luck. The most stark illustration of this was the explosion on the Deepwater Horizon drilling rig in the Gulf of Mexico. Just about the time of the blast, which resulted in the loss of 11 lives and the biggest oil spill in history, management were visiting the rig to congratulate the team on their safety performance, having previously met their zero-harm targets.

Managing risk appetite

The reality is that everybody has their own risk appetite. And for most people, this does not equate to an absolute intolerance for any injury at any time, ever. We need to understand the risk appetite of our workers and work with them through that lens. Allow them some leeway to make their own decisions where things are familiar and generally low-risk, as well as providing input in resolving higher-risk issues. If we disengage them in the routine, they will not be in the right mind-set to think about the non-routine, and that is where the true danger lurks.

To further complicate matters, research has shown a phenomenon known as risk homeostasis, or risk compensation (I should note, for balance, that there are some critics of this theory out there). The theory is that when people feel safer, they take more risks, subconsciously maintaining the risk at a base level with which they are comfortable. Much of this research has been in the road traffic area. Studies have shown that if people wear seatbelts to reduce risk, they drive faster and increase it again. If road markings are clear and visible, people will drive faster as they are more confident of knowing where the side of the road is. This process works both ways. When pedestrians and vehicles are not clearly separated in shared areas, increasing the risk of pedestrian injury, people drive more slowly and carefully, reducing the risk. Where such compensation occurs, trying to eliminate all injuries by ever-increasing risk reduction appears to run counter to human nature.

The focus on numbers that zero harm drives is the same as the triangular fallacy. By reducing numbers instead of reducing risk, it actively works against preventing serious injuries and fatalities. The intent is laudable and sincere, but the consequences are the opposite of its intent.

People naturally accept and understand a certain level of risk. The only way to guarantee prevention of all injuries is to remove all risk. If you ask anybody if it is possible, logical or desirable to remove all risks from life or from work, they will reply in the negative. Why, then, do so many businesses seem to think it is a sensible approach to try?

Key points

From this…

- Zero harm drives incident counting.

- Zero harm rewards lagging output.

- Zero harm has an emotional basis.

- Zero harm is a demotivating target for staff and a vague basis for improvement.

… to this

- A risk-aware approach focuses on those areas that have the highest risk potential.

- Improved inputs into safety processes are rewarded.

- Safety management has an objective and rational basis to drive improvements.

- Realistic targets are set based on specific improvement areas.

12 The worker implication

Where do most of our opportunities to improve come from in safety? The vast majority of organisations don't have the time or resources to commit to research, trials and carefully considered experimentation of different options. The very fact that we still base much of our safety process on decades-old research demonstrates this extremely well. So, how do we improve?

For most of us, we learn from our mistakes. In safety terms, this means learning from accidents. In the better companies we will also learn from others' mistakes – via safety alerts and the like – and also from near-miss incidents. But it boils down principally to incident investigations (we will look at other places to learn from later on).

Just an aside while we are on the subject of a near miss. There has been a tendency among some safety professionals to term it a 'near hit', citing the need to point out the fact that it is important to emphasise the word 'hit' just in case people fail to recognise the significance. As if 'miss' is somehow insufficiently threatening. As if the workforce can't comprehend that a miss can be important. This is another of those patronising 'we know best' attitudes that the safety industry seems so good at. While the use of correct language is significant and 'hit' rather than 'miss' may invoke the right response in isolation, the reality is that 'near miss' as a phrase means something that is worryingly close. People interpret it as such and not as, 'That missed so we're all okay.' Furthermore, 'near hit' is a clumsy, obviously fabricated phrase that is not part of normal language. It feels contrived and uncomfortable – hardly a way to embed the concept in our routine. If you genuinely feel that 'miss' conjures up the wrong picture, at least change it to something that is easy to use in practice. Perhaps 'close call' conveys the right level of urgency, while remaining a normal part of the language.

Anyway, back to investigations. Learning from investigations is important and there is much that is useful to be learned. But, in order to do so, the investigation needs to be effective in identifying the shortfalls and causes that require correction. Given all the shortfalls in safety thinking we have identified already, this is tricky enough to get right, but during an investigation all of this is swamped by one overriding factor that undermines the whole approach.

It's the worker's fault.

So many investigations come up with this finding. This is particularly the case in simple shop floor accidents, such as those in the residential and light commercial construction industry, but is also repeated throughout supposedly more mature industries, albeit better dressed in fancy reports and root cause analysis tools. In fact, even in the aviation industry, otherwise one of the most advanced and forward-looking from a safety perspective, the vast majority of crash investigations still cite 'pilot error' as the principal cause.

A former colleague refers to these as 'man dead, own fault' reports.

I once had a conversation with a CEO not long after a contractor retained by his business lost his foot after using it to push foliage into a wood chipper. It was a common cry from the CEO: 'Whatever happened to personal responsibility? How do we stop people doing things that are so obviously unsafe?'

Or, to quote a draft regulatory guide I once read that fortunately got changed before publication, 'Why do smart people do stupid things?'

The injured party was only young and inexperienced, so I answered the CEO's question with one of my own. What was it that had been going on in the workplace that made him think that carrying out such an obviously unsafe act was reasonable? Had he seen others do it? Was he under pressure to work faster? Had he been working too many long shifts?

The first response in many of these situations is to wonder about the sanity of the individual. We typically look at the situation with a sense of superiority: 'I would never do that. What an idiot.'

But if we truly want to understand what happened and why, in order to prevent similar events, we have to look behind the action to the reasons for it. We don't learn by simply stating that the person was stupid and shouldn't have done it. In fact, we obviously don't learn because we, apparently, continue to employ stupid people. It's strange how they seemed fine when we hired them, but now it turns out they're stupid.

It is quite commonplace to see a 'safety fail' meme on internet forums followed by numerous smug safety people commenting on the idiocy of the people in the picture. Stupid people convicted and sentenced to ridicule without trial or opportunity to defend themselves.

In short, if your investigation concludes with, 'Worker A failed to…' or 'Worker B did not follow the procedure to…' or 'Worker C did…' then it has not gone far enough to understand why.

I have a great deal of sympathy for the construction supervisor who is required to investigate an event on a site and who has received little or no investigation training. I have no sympathy for safety professionals doing it or allowing it to continue in their organisation. There are two principal factors that I see underpinning this failure. The first is a return to our friend Heinrich. The second is a failure to understand how people make decisions.

Eighty-eight per cent of accidents are caused by worker behaviour

This is another favourite based on Heinrich's work that has entered safety lore. Again, this is not to belittle Heinrich, who was pioneering in this area, but rather to berate ourselves as a profession for not moving on. As part of his research Heinrich came to this conclusion by looking at the stated causes of accidents. At the time, there were no advanced investigation methodologies, no understanding of systems thinking and no sound knowledge of human decision-making processes, so this is not an unreasonable finding to make. But, as with his triangular statistics, by now it should have been confined to the annals of history.

Let's be clear. Almost every event will have some form of action by an individual that will be crucial to the outcome and, with hindsight, probably inappropriate. Most of these will be front-line workers (maybe even 88 per cent). But the worker is simply the last link in the chain, the sharp end of the instrument being wielded by the organisation. Workers do not take a perfectly developed and designed fail-safe process and implement it. They typically fight against resource shortages, poor procedures, lack of training, unexpected circumstances and do the best that they can under a number of competing and conflicting pressures. In most circumstance it is a miracle that they get anything right at all.

Blaming the worker for an accident is like blaming a soldier in a warzone for a death, while ignoring the superior officers giving the orders, the politicians going to war and the stress of the whole situation. The soldier may have pulled the trigger, but the situation was not of their making.

Yet, we continue to do this. Whole industries around 'behavioural safety' have sprung up that, even if it was never the intent, have turned into blame-the-worker processes. They take a limited consideration of human behavioural science and attempt to fix the symptoms rather than the cause, defining unsafe behaviours that are 'bad' such as rushing, inattention and so on without determining the underlying reasons for them. At best, this becomes a band-aid solution that may stop individual failures, but never the systemic ones. At worst, it degenerates into bad feeling and persecution.

It is important that workers are well trained in recognising behavioural symptoms so that they can avoid them becoming unsafe acts, but we can't simply stop there. If we recognise that people are rushing, what does that say about our planning and project management? Does it mean we have insufficient staff for the workload? Have we not got sufficiently experienced workers who can do the job quickly while maintaining accuracy? Is our 'just in time' delivery system actually 'just too late'?

At its core, the 88 per cent quote is no more useful than saying, 'Most of our activities involve people.' We need to recognise that the worker is simply the end point in a supply chain for safety, all of which needs to be considered in accident investigations. Only then will we genuinely improve as much as we can.

Note that this is not to say that individual workers are never culpable. We just need to make sure that we have rigorously and thoroughly examined all aspects before coming to our conclusion. It is also worth noting that it is very difficult to learn from an incident if we have just fired the people involved. Very, very rarely does a worker do something that is deliberately wilful and malicious.

Decision making

The other difficulty in undertaking an investigation when someone has apparently acted unsafely is centred on decision making and our understanding of it. Some of the best information relating to this topic can be found in the excellent book *Thinking, Fast and Slow* by Nobel Prize winner, Daniel Kahneman (2011).

It is often difficult to comprehend how someone can do something when it seems so obvious that it would end badly. The first component to this is hindsight bias. When we investigate, we do so in the knowledge of what actually transpired. This makes it very easy to see why it was a bad idea, as all the evidence points to it being so. At the time the decision to act was made, however, the outcome was not certain. The person did not know they would have an accident, especially if this was something they had done before without injury. Hindsight bias makes us overestimate how obvious the outcome was. How often do you hear people state after an event has occurred that they had always thought that would happen? Yet, it is difficult to remember them highlighting it as a risk in advance. This is hindsight bias in action. It is also supported by post-event rationalisation. The human brain, and its memory, is a very tricky beast. After an event we will reorganise and change our memories to make them fit with our rational view of what we think we should have said or done (which is another point to note in investigations when interviewing witnesses – memories are extremely unreliable).

When we look back over why a decision was made, we tend to imagine a logical, thought-through process that weighs up all the relevant factors. This is reinforced in safety by our use of risk matrices, decision trees, job safety analyses and so on. But the reality is quite different. In life we make hundreds, if not thousands, of decisions every day. If we stopped and thought through them all logically and weighed up all the pros and cons and possible outcomes, we would be paralysed into inactivity. Thinking is hard work, so the less we can do of it, the better. Ask someone to solve a tricky maths puzzle while walking and they will stop. Thinking hard *and* walking is simply too much for the brain to cope with.

To overcome this, we use a combination of previous experience, assumptions, values, biases, knowledge, constraints and pressures (and others) to develop some broad decision-making rules, or heuristics. A heuristic is essentially a rough and ready, practical decision-making process that is not optimal, but is quick and that our experience has shown us tends to work well enough. It is a trial-and-error approach or a rule of thumb. But we don't just use it for things we have done before and learned from. We will also use it for decisions that are reasonably similar to previous ones. We will even replace a difficult question or decision with an easier one that we know the answer to. Ask someone a more simple maths question while walking and they can answer without breaking stride. They don't think about it too hard, simply pulling up all their previous experience (learning times tables at school, for example) and producing the answer.

We do not use heuristics for everything. Some decisions we recognise as important enough to slow down and think through.

For the majority of cases when we make a decision at work, we are in the fast, heuristic mode. We are not usually consciously weighing up the risks, but broadly considering them in light of what we've done before and our knowledge of current situational factors. Most people have a reasonably well-developed appreciation of risk versus benefit for the majority of situations that they come across, developed through years of life experience. As previously mentioned, most would take a significant personal risk to save a child, but not such a risk to save a dog. But in an emergency situation requiring action to make the rescue, this would be a split-second heuristic decision based on experience, not a carefully thought-through, logical choice. It works in a broad, most-of-the-time sort of way in circumstances we recognise. When a situation that is less familiar or more complex appears, the logical muscles fail through lack of use and the heuristic ones take a guess, often getting it wrong via one or more of a whole multitude of biases that we all have.

In understanding why workers take the actions that they do, understanding their decision-making process is vital. This will help us pinpoint the real areas of difficulty and make our processes forgiving of the reality of human nature in decision making.

Most of traditional safety is based on an idea that our systems and processes are safe and work well, but the workforce sometimes fails to follow them and that is when an accident happens. This is a fundamentally flawed perspective that we will explore further in a later chapter.

Key points

From this ...	... to this
• Investigations are focused on what the worker did wrong.	• Investigations go deep enough to understand the context for worker actions.
• Workers are blamed for accidents.	• The role of the overall system is recognised in accident causation.
• Poor decision making is punished.	• The reality of human decision-making processes is understood and applied to understand why decisions were made.
• The failure was obvious.	• Hindsight bias is recognised and local factors explored when understanding why failures occurred.

Reference

Kahneman, Daniel (2011) *Thinking, Fast and Slow*, Farrar, Straus and Giroux.

13 The safety separation

When asked what their personal version of safety utopia is, the majority of safety professionals will choose some variant of the following:

> The perfect future is one where I don't have a role because everyone builds safety into their normal jobs and just does it as a matter of course. There is no need for a safety adviser/manager.

This is fairly unlikely – most people think about profitability and costs all the time at work but we still have armies of accountants and a CFO – but it also seems to be something that we actively work against rather than towards. Perhaps this is because nobody really wants to work themselves out of a job, even if only subconsciously. Or perhaps it's because few of us actually pay any attention to what we say.

Considering the ubiquity of this backdrop of 'building safety into our normal jobs', we spend an inordinate amount of time treating it as something different.

Safety as something special

We have safety moments at the start of meetings; we have safety days and safety stand-downs; we put safety first on our meeting agendas; we have safety posters plastered around our offices and work-sites; we have leaders engaging in safety conversations. The list goes on and on. What is the consequence of this? We isolate safety from the general workflow and turn it into something packaged that we only think about when it is put in front of us and pointed to, which is the exact opposite of our stated vision of perfection.

Next time you sit in a meeting where a safety moment is presented at the start, pay attention not to the content, but to the other people in the room. Watch the level of engagement and see how well it is working to keep safety at the forefront of people's minds (more on this later in Chapter 15).

Every time we separate safety out and treat it as something different, we expose ourselves to the danger that the way we are portraying it is misaligned with our audience – something we are very good at, as we have discussed. Every time we do this, we drive the wedge between safety and operations a little bit deeper. This is not helped by the fact that these separations of safety are almost universally 'tell' sessions rather than 'ask' sessions. Even where we do ask for feedback, it tends to be based around a chosen theme and contrived to move towards the answer we expect to see.

The unspoken and underlying message in all of this is one of: 'That's the safety requirement out of the way – now we can get on and talk about the important stuff.'

We don't do these things for any other function or activity, so why do we do it for safety?

Perhaps it's because safety is our number one priority, but we have already dispelled that particular myth. The main reason stated is to keep a high focus on safety, keeping it 'front of mind'. This requirement to keep focus high in order to keep people safe is rooted in that most pervasive and pernicious trait of safety management – blaming the worker.

We touched on this in the previous chapter in relation to specific incidents and their investigations, but this is where it really starts to take hold and shape how we operate. So many incidents when viewed in hindsight involve a failure to take an action that seems obvious, or carrying out an activity that was obviously foolish. The inevitable conclusion of these is that the worker was complacent, careless, inattentive or something similar. Over time we aggregate these into a conclusion that workers are not paying attention; therefore, they must not be thinking about safety. It then becomes our duty to remind them about it at every possible opportunity, leading inevitably to the safety separation.

This is symptomatic of one of the principal problems of safety management. Most businesses subscribe to the paradigm that, as organisations, they know how to do the job safely, if only those annoying workers would start doing it properly and stop screwing it up. These typically fall into two groups. One is the 'mature' organisation that has written procedures and processes for everything that purport to manage all foreseeable risks. The second is the less mature business who thinks that there is no need for procedures because it is 'just common sense'. These two points of view are radically different in their application, but they converge and are fully aligned when it comes to laying all their problems at the feet of the poor worker. Whether they are failing to follow procedures or failing to apply their common sense, it is still their fault.

But instead of fixing the problem, the safety separation exacerbates it. We separate safety out, couch it in terms that don't resonate with the workforce, focus on risks that they don't think are important and then design systems that are not accommodating of the reality of the way people actually think and behave. All of which combine to frustrate, divert and generally complicate matters, making accidents more likely that will then be investigated and the conclusion drawn that the worker at fault was insufficiently focused on safety, and the whole sorry process starts over again.

An often repeated definition of insanity is doing the same thing over and over again while expecting a different outcome. But this is our approach to improving safety. We'll try to refresh and repackage the message in a half-hearted recognition that we should shake things up a little, but, fundamentally, whether through posters, moments, walkarounds or stand-downs, the most tried and tested (although sadly not proven effective) approach to safety improvement is to tell our workers to be careful. Over and over and over again.

This approach also reflects an inability to recognise the complexity of safety. It is not generally helpful to expect simple solutions to solve complex problems, but the 'be careful' school of safety management is doing just that.

'Safety is as simple as ABC. Always Be Careful.'

Really?

I have seen this on a safety poster. Not only is it a ridiculous over-simplification, it is incredibly patronising. Presumably someone commissioned this poster and then designed, reviewed and approved it for publication. What did that thought process consist of?

> 'We've been having accidents on site and we want to reduce them.'
> 'Do we know why?'
> 'Well, our investigations have shown that people aren't paying attention.'
> 'Okay, so they need reminded to be careful.'
> 'Hey that's great. We can tell them to be more careful – I'll bet they've never thought of that.'
> 'Well, obviously not; otherwise they wouldn't be having accidents.'
> 'That's agreed then. Let's turn being careful into a pithy slogan and put some posters up.'

Not only does it patronise those workers that we employed because, presumably, we thought they were smart and capable, but it also tells anyone who has had an accident that it must have been their fault because, after all, safety is simple, isn't it? And so we come full-circle back to blaming them again.

Almost without exception, all safety posters (and related slogans) work in the same way. They're not informative, they're not smart and they lead inevitably to the same conclusion. Yet there is a genuine 'manage by posters' mentality out there. Had a number of vehicle accidents? Put up a poster reminding people not to crash the car. Seen a spate of hand injuries? Flood the workplace with posters about the line of fire.

A similar blame failure can be applied to the 'responsibility' messages that are also commonly found. A particular favourite bad example is the statement above the mirror that says: 'This person is responsible for your safety today.' While recognising the positive intent, what does it say to the person who has an accident? Our message is that it is their responsibility and that, furthermore, discharging that responsibility is really easy and just takes being careful. We might as well put up a poster that says: 'Accidents. They're Your Fault, Stupid!'

Whose responsibility is it anyway?

Imagine an executive team sitting in their boardroom considering the latest profitability figures. Sales are down, costs are up and profit is (obviously) being squeezed between the two. The signs are worrying. What do they do about it? They engage with the sales and marketing team to increase revenue; they review the operations to understand where the increased costs are coming from; they assess the potential for overhead reductions in the corporate office; they consider potential inefficiencies that could be fixed in maintenance, in procurement, in manufacturing, warehousing and distribution. In short, they engage with the operational management across the organisation and work out how to fix it. They don't declare it a money problem and get the accounts team in to provide the answers.

Yet this is exactly what happens if it's a safety issue. When they have poor safety performance, they call in the safety manager and ask them to fix it. The same thing happens at operational levels – we've had an accident, get the safety adviser in to investigate.

Safety is an operational and management responsibility. It is not the responsibility of the safety manager or the safety team. This may sound like an abrogation of responsibility, but the reality is that it is line management's job to provide a safe working environment. The safety specialist is there to be a trusted adviser; to provide expert advice on the more complex aspects and to take a strategic and independent overview of the whole process. They are there as custodian and support, in exactly the same way as the CFO and accounts team provide expert tax advice, statistics and annual overviews but they are not responsible for day-to-day cost management and sales within the business.

The safety adviser can only spend so much time with the working team. The only way for safety to be fully embedded within normal operations is for local supervisors and managers to take full responsibility for the day-to-day routine aspects of it. Stop separating safety out and start building it in. However, in order to do this, we need to spend less time telling people what to do and more time listening to them to understand the issues and play a supporting role in managing them safely. We will explore this in more detail in Part III. We can also make a start by stopping drowning them in bureaucracy, which we will explore in the next chapter.

Key points

From this…	…to this
• Safety is separated out from normal operations and treated as an add-on.	• Safety is a routine part of day-to-day operational thinking.
• Safety is looked after by the safety manager.	• Safety is owned by line management with enabling support from the safety team.
• Posters and slogans remind people about safety to keep their focus up.	• Workers don't need additional reminders about safety, because it is a natural part of what they do every day.

14 The systemectomy

There are many problems with safety legislation, but there is one fundamental irony that leads to significant problems within the safety world. As with many other problems discussed in this book, it is centred on the worker.

In much safety legislation around the world, there is a requirement to consult with the workforce. They have a right to appoint health and safety representatives, have health and safety committees and be consulted by management over decisions related to the management of significant hazards within the workplace. When industry task forces, review bodies or commissions of enquiry are established, there is a strong drive for them to be tri-partite in composition – that is, comprised government, industry and unions.

The rationale for this is twofold. First, members of the workforce are the ones predominately being put at risk. It is only right that they are consulted adequately around the issues. There is an obligation to properly inform and collaborate, given that they face the maximum exposure. It is the safety version of 'no taxation without representation'. Second, there is recognition that, in many instances, those exposed to the hazards on a regular basis have many of the best ideas of how to manage them. In our workforce we have an army of experienced, knowledgeable and practical practitioners with first-hand insights into what works, what doesn't and what factors are involved at the sharp end of the operation. This knowledge should be harnessed and used to make improvements.

Within reason, this is correct. Obviously, not every individual can add significant value, but as a collective, there is real merit in this involvement and consultation process. The shame is that it needs to be legislated to make sure all organisations do it.

The irony arises when other parts of legislation are considered. These are the parts that require companies to treat all employees as idiots incapable of independent thought. I'm paraphrasing somewhat – there is little legislation that actually uses the word 'idiot'. But, when combined with case law into the outcome of safety prosecutions, the idiocy component is broadcast loud and clear.

Legislation varies from country to country. In some jurisdictions (such as the US) it is very prescriptive, clearly specifying the activities that must be carried out to be safe. This is the ultimate in idiot control. I will tell you what to do in any given situation. When an injury or death occurs, I will look into it until I find which part you failed to do and then I will prosecute you.

In other places, so-called target or goal-setting systems are in force (e.g. in the UK). These set up frameworks that require the organisation to control the risk of injury to a reasonable level as their target, but are fairly open about the means to achieve that. It is incumbent on those in control of the hazards to identify them, inform people about them and put adequate control measures in place. In my view, this is a far superior system, but does require smarter and more sophisticated regulatory management, which unfortunately is not always the case.

In either system, the outcome of court cases tends to be the same. The failing is almost always identified as within the safety procedure. You failed to identify the hazard. You failed to inform the person of the hazard. You did not adequately guard the machine. You failed to train the person to use their equipment properly. However, none of these cut to the root causes of cultural understanding of safety within organisations. Few (if any) prosecutions occur where the culprit is identified as a failure to develop a learning culture within the organisation, or where the company inappropriately fostered a culture where the pressure to progress was so great that people were prepared to make safety shortcuts. In fairness to the regulator, these are difficult cases to prove versus whether or not there is a record of training on a particular piece of equipment.

But, when someone loses a hand because they put it inside a live machine to free an obstruction, surely the fundamental question is not whether the guarding was adequate (although this is a valid question, just not the most important one), but why on earth they felt it was okay to take that action. Note that I'm not blaming the individual here. People do things for a whole variety of reasons that look obviously wrong with the benefit of hindsight bias. Many of them relate to external pressures that should never be imposed. But the key thing is that improving the guarding will only prevent that exact same incident occurring on the same machine. Changing the organisation's understanding of risk and improving the culture to one of mindfulness, learning and the courage to slow down to be safe will also prevent many other incidents occurring.

The first activity in almost any investigation is to look at the procedure and see if it was followed. But rarely is the question 'why' adequately considered.

Of course, we covered all of this in a previous chapter. The question here is less about the internal investigation and more as to why the regulator (actually, probably more the prosecutor) seems unable to properly ask why.

Interestingly, when investigations are commissioned into large-scale accidents – by government enquiries, royal commissions, industry bodies, etc. – the role of culture is often very clearly identified and highlighted as one of the key causes, often the most influential cause. In this instance, where there is no legally imposed burden of proof, it is much easier to identify culture and leadership as an important factor. But these are often intangible and difficult to demonstrate in any clear-cut way. Investigations into Pike River, Piper Alpha, Deepwater Horizon and others all supported the need for improved cultures.

By taking the enforcement option, the mind-set is developed that evidence of compliance becomes key. A plethora of systems is developed to ensure that training is recorded. All of the hazards are written down in a register, so we can demonstrate that they have been communicated. Everyone signs on to the hazard assessment at the start of the job, so it can be shown that they knew what the hazards and controls were. Although nobody listened, read or thought about any of these things before signing, because the poor culture remains. The workforce, who have a morbid hatred of paperwork at the best of times, sees this for the arse-covering exercise that it is and responds with a 'tick and flick' mentality of form filling. Let's get the toolbox talk out of the way so we can get on with the job.

A clear demonstration of this occurs within many inductions. One of the difficulties in an induction is knowing whether or not the trainees (or victims, given many inductions I've seen) have understood and absorbed the information. The answer generally is that they haven't, because there is usually far too much to remember, delivered in a tedious fashion and, for contract and itinerant workers, it may well be one of a dozen or more sites where they work, each one with its own requirements. However, we rarely see companies collaborating to develop a single base induction with some local variations, or carefully constructed progressive inductions that apply over time with the help of an experienced 'buddy'. No, what we see is a questionnaire at the end to 'demonstrate' the learning received. It's no good for everyone to fail it, so it becomes a ridiculously simple multiple choice list that can be completed either during the course as each issue is covered, or in discussion with the presenter at the end.

A typical question may be:

If you see someone carrying out an unsafe act, do you:

(a) Pretend not to notice and carry on – it's none of your business.
(b) Intervene and stop the job to ensure it can be done safely.
(c) Applaud their courage.
(d) Don't say anything because the person is your supervisor.

This may look ridiculous, but it is actually fairly close to the truth in many workplaces. This is a classic example of the system being developed for compliance purposes by the safetycrat. It adds zero value – in fact, it is probably harmful given people's response to it – but it fulfils a need to *prove* that something was done. This is symptomatic of many safety processes in place today.

What is required is a *systemectomy*. Get rid of 90 per cent of the paperwork and let's train our people properly and develop appropriate cultures. Human behaviour is a highly complex, multi-dimensional problem comprising different viewpoints and lenses through which people see the world in a myriad of different ways. The workplace is a complex, varying environment that changes in unforeseen ways every day. Attempting to manage this entire spectrum through one-size-fits-all pre-determined paperwork systems is simply not going to be effective. But this is what we see. An all-encompassing written system that must be complied with.

Compliance

The traditional role of a safety professional is seen as one of compliance. The safety police enforcing the rules. Upbraiding people for not wearing the correct safety gear. Even the terminology is often compliance-centric – an authoritarian-sounding safety *officer*, although there are now many safety *advisers* out there. I recently came across a job advert for a 'safety *compliance* officer'. What message does that send to the workforce when you introduce yourself on the first day? It's hardly going to foster an open and productive dialogue to move forward together.

This compliance theme spreads throughout the organisation. A number of companies now have a set of rules in place – golden rules, life-saving rules, fundamental rules or similar – to save the idiots from themselves. My first response to these when I heard them was how patronising they were. They are all made with good intent. These are the things that people do wrong when they get killed in industrial accidents, these are the things we must stop – and so rules are developed. The problem is that telling someone not to work on live equipment will not save their lives. They didn't make the decision to expose themselves to the risk because they didn't know about it. It's incredibly obvious to anyone with the relevant work experience. The question is why they did it *in spite of that knowledge*. Just reiterating the danger is pointless.

In fact, it is worse than pointless, it's actually detrimental. The workers feel patronised. They have worked on the rigs/sites/plant for 20 years and someone comes into a classroom and tells them not to work on live equipment without isolation. Stating what is painfully obvious to anyone as if they are a child who knows no better simply results in them losing all respect for safety in its corporate guise, setting the scene for future problems.

It is akin to telling smokers to give up because it is bad for their health, or obese people to cut down on fatty and sugary foods. There is an oft-repeated mantra in the media that this is all about education to help people understand the implications, but this is nonsense. Any of these people would have had to have lived in a cave for their whole life to not know the dangers of smoking or over-eating (or drug use, or alcohol abuse or many other similar problems). It is not a lack of knowledge that is the issue.

People's behaviours are complex. One model talks about ABCs – Antecedents, Behaviours and Consequences. This theory states that the behaviours we see are preceded by certain attitudes, conventions and contexts that drive behaviour in a certain way. These are the antecedents. They may be cultural and deep-rooted, or they may be very context-specific.

Behaviours are also modified by what is known about the likely consequence. Significant and immediate consequences will drive behaviour much more strongly than small or distant ones. This is part of the rationale behind smoking. The negative consequence (long term, stochastic cancer death) is outweighed by the positive one (immediate and definite pleasure from the nicotine hit). If a smoker was given only cigarettes laced with arsenic that would be guaranteed to kill them quickly, would they give up? Probably, although maybe not definitely given the strength of the addiction, but they would probably not have started in the first place. Although, even that is uncertain. There are still people trying methamphetamine despite the widely known fact that it is both incredibly addictive and frequently lethal in relatively short order.

Part of the problem with unsafe activities is that the consequence is seen as distant with a low degree of probability that is worth risking – especially if it is a shortcut that has been successfully taken before. The intent of the rule is to bring more immediate and definite consequences to bear, which is theoretically sensible, but ignores the complexity of dealing with people.

The systems we put in place are built around this compliance mentality.

Procedural lobotomy

Putting ever more rules and systems around these areas is dumbing down the safety process. It is an over-simplification of the issue and will not make lasting change. Lazy thinking, not critical thinking.

The many facets of this issue are well covered by Long and Long (2012), drawing on psychology research into thinking and learning modes.

The need for compliance also drives the need to develop procedures for all activities. This leads to two key factors that are detrimental to safety. First, a longer-term position where all initiative is removed from the workforce; and second, a lack of understanding and recognition of what activities are safety critical.

In an environment in which there is a high emphasis on compliance at all costs, workers become reliant upon the procedures to tell them what to do. Any variation from the procedure is punished, reinforcing the compliance requirement in a vicious circle. Over time, when a situation arises that the procedure does not cover, the worker stops and waits for an analysis of the situation by someone else (who tends not to know the elements of the task as thoroughly). Once the analysis is complete, the procedure is updated and the job can continue. There is an obvious efficiency problem here as production is stopped (but safety comes first, right?), but the main concern is that the worker loses the capability of responding. This procedural emasculation drives rote performance and lack of initiative. It lobotomises the organisation, removing innovation, entrepreneurialism and responsiveness. When an unusual situation occurs, workers no longer have the capacity to respond unilaterally.

When most significant injuries or events occur, particularly major accidents such as explosions, there tends to be a rapid escalation of unpredicted combinations of events. When this happens, there is no time to convene a committee to review and determine what needs to be done. If the incident is to be avoided, there has to be an experienced, competent workforce capable of assessing and responding in quick time – even if this is limited to recognising their inability to manage the situation, prompting evacuation to limit the consequence.

Responding to minor issues and challenges provides a degree of capability to respond to more significant ones. Over-reliance on procedures destroys this capability. Emergency exercises can provide some assistance in this area, but these tend to be somewhat contrived and do not have the urgency of the real thing. There is no substitute for experience.

This is not to say that nothing should be proceduralised. There is a need for checklists and guidelines to assist even the most seasoned individuals and the more complex a plant or process, the more the requirement for clear procedures. But if absolutely everything is bound by rules, how does the workforce know which operations are genuinely high risk? If everything is important, then nothing is important. There is a need for a graduated scale, with clear understanding combined with rules and procedures around those activities that are safety critical and freedom to operate based on experience and knowledge for more minor ones.

In a nuclear facility dealing with plutonium, for example, criticalities can occur when too much fissile material accumulates in one place. This essentially replicates what happens inside a nuclear reactor, but in a location that is not designed to contain it. The resulting radiation is undetectable (by people) but deadly. In the unlikely event that this occurs, the safest evacuation route is to avoid this location, unlike most evacuations where the preferred route out is the quickest one from the building. Evacuation routes are carefully designed to avoid any areas where fissile material is held and therefore have the potential for criticalities. In many cases, this route is counter-intuitive and it is, therefore, vital that clear procedures are laid out with no room for initiative and opinions of individuals. Developing criticality models is a highly specialised field – somewhat of a black art even to most nuclear workers – so this is something that must be handled by careful consideration and with reference to the criticality safety team. Emergencies are still practised, but only to make sure the processes work and people follow them properly.

However, if the workplace is swamped with dozens of rules that must not be broken, even when some of them are clearly not a high risk, the integrity of the rules is lost and people will question their validity. They may then not obey the ones that genuinely do matter. One 'golden rule' may state that seatbelts must be worn at all times. The intention is good, because seatbelts can, and do, save lives. But on most industrial sites, vehicles cannot get to a speed where lives can be threatened. Because it is a rule, it must be enforced and there is no room for discretion. I have seen people disciplined for not wearing a seatbelt when reversing a vehicle at very low speeds because they broke a rule, although there was virtually no risk.

Rules must only be set where there is genuine high risk and/or high complexity. Again, it is not the intent that is the problem, but the unintended consequence that has not been thought through. Worse still, many organisations roll out these rules with an approach that one worker described as 'comply or die', with the consequence being dismissal. In the seatbelt example, although the individual was disciplined, he was not dismissed – and rightly so. But once an absolute standard is set it must be complied with. When the rules are not properly thought through, either the stated punitive action must be taken when it is ridiculously over the top, or the action is not taken as promised, because it is too extreme for the situation. In either case credibility is lost. I know of someone who was dismissed because they used their electronic access card to allow someone else through a locked door against the rules. Their crime was letting another staff member into the canteen.

The use of absolutism in setting the rules resulted in having to take action that was completely out of all proportion to the event. The rationale was given of, 'If we can't trust people to comply with minor rules, how can we trust them to comply with more important ones?' But this is nonsense. It is akin to expecting someone caught speeding to be equally capable of murder.

People respond better to being cared for than being told what to do. A fatal risk programme that identifies major risks and provides information to help workers make better risk-informed decisions will contain essentially the same information as a set of rules. But much more buy-in is achieved from a message that says, 'We care about your wellbeing – please remember these principles, it could save your life' rather than 'Don't do this or we'll fire you.'

Note that there are good behavioural reasons sometimes for imposing blanket rules. Seatbelts are actually a good example, because their regular use becomes a habit, so making it more likely they will be used when there is a higher risk involved. Similar arguments can be made around the wearing of protective equipment at all times. However, these need to be carefully thought through and carefully implemented in a way that recognises and explains why it is being done, with appropriate responses if not carried out. In most cases, there is complexity. Gloves prevent cuts, but can actually cause injuries due to getting caught in equipment, or by reducing dexterity. In almost all circumstances, imposition of general rules is counter-productive as there is always an exception. This is what the workforce will focus on and what will disengage them. It is human nature to look at the exception in isolation.

So do we need rules at all? Clearly it is unreasonable to expect workers to make every single decision based on a detailed risk assessment of the circumstances at hand. In these instances rules can be helpful to provide a rapid solution. Yet there are other occasions where existence of the rule implies safety right up to the point where the rule is breached, when this may not be the case if there are additional risk factors involved.

Rules are most beneficial when not all the information is known (or even knowable) and people cannot make risk-informed decisions. This is typically the case in complex systems with high hazard potential where a quality decision can only be reached by careful consideration of all factors by a multidisciplinary team pooling their knowledge. Staying in our nuclear waste facility, there is a safe upper limit for the surface density of an array of stored material. It is not possible for a process operator to make risk-informed real-time decisions about the structure of the array, but (in combination with other controls elsewhere) a simple rule can be established about the number of containers allowed in a stack.

Limiting rules to certain circumstances where risk is higher has the benefit of emphasising the importance and so making compliance more likely. They must also be quite specific. The broader the rule – always wear a seatbelt – the more likely it is to be seen as inappropriate in some circumstances and therefore optional. After all, most of us have broken the speed limit because we know it isn't realistic in all situations and there are times when it can be broken without high likelihood of accident (with apologies to all traffic police).

Hopkins (2010) has implied that the control pendulum has swung too far towards risk management and needs to swing back to more rule-compliance, arguing that operationally workers need the simplicity provided by rules. In fact, in the absence of specific rules, in some circumstances they will make their own rule of thumb to follow and help them make decisions. He does, however, recognise the need for balance between the two. As ever, there is no black and white, right or wrong in safety. How do we find the balance in this grey zone?

- Be careful of phrasing. Couch requirements in terms of supporting safe action, not in the language of absolutes and threats.
- Impose rules only where risk is high, to emphasise their importance.
- Impose rules for specific, usually complex situations, where local decision making is difficult.
- Use the rules to build a framework within which workers are given the licence to use their core skills to change, adapt and improve.

When the framework becomes challenged or changed, involve the workers in consideration of the implications. A clear boundary is established outside of which the risks are too great, but inside it workers have the autonomy and core skills to manage their worksite and its safety as appropriate (Figure 14.1). This autonomy is extremely motivating and engaging for the workforce.

This is not to say that there is no operational guidance within the low-risk area, but this can be provided by checklists to ensure operational steps are not missed that may have production implications, or even minor safety implications.

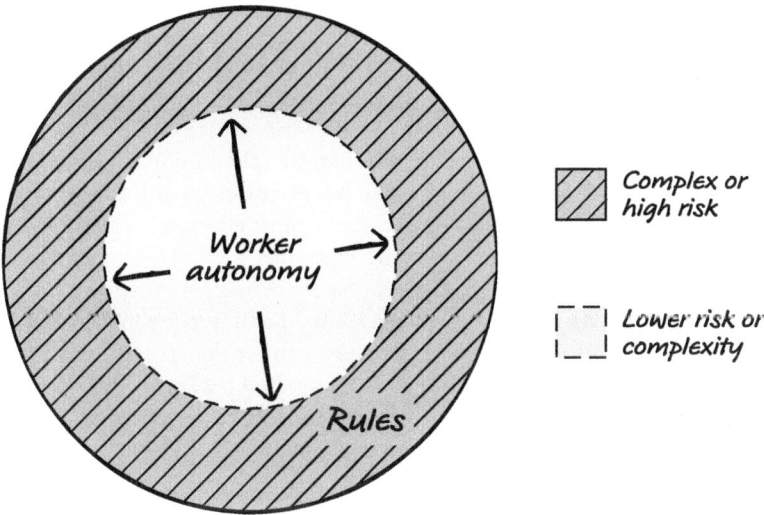

Figure 14.1 Rule–autonomy framework.

Rules are the power tools of safety. They are labour-saving devices that do most of the work, but have their own significant risks if mishandled and fail when it comes to the precision needed for a fine finish. For that we need to overlay the hand tools of carefully applied risk management. It's slower, it takes more focus and more expertise, but it can achieve that final few per cent of improvement.

A key part of the systemectomy is to remove those rules that do not actually add to the level of genuine risk reduction. Rules are helpful in certain, specific circumstances where risks are high and the appropriate action to take is clearly defined and unambiguous.

System growth

As changes occur within an organisation, whether due to operational changes or due to incidents occurring, the safety management system typically grows to encompass them. Rarely does anything get removed from a system. Coupled with poor-quality investigations frequently citing the need for more awareness and more procedures, this leads to compounding growth as changes accrue year after year until the whole system is so bloated and bureaucratic that it becomes completely unwieldy.

This turns people off. They don't know where to find useful information buried beneath all of the superfluous documentation, and eventually the system does more harm than good. The systemectomy should be applied via a re-examination of the whole purpose of the system (see Part III) or at least by questioning the real value being provided by each component.

Key points

From this...	... to this
• Compliance is the basis of the safety management system.	• Rules are retained only where work is high-risk and complex.
• Activities are only done in accordance with a procedure.	• The skill and experience of workers are relied upon to undertake lower-risk tasks in the way which is most effective in the circumstances.
• The safety management system is bloated and bureaucratic.	• The safety management system is streamlined and designed to enable processes rather than stifle them.
• Procedures are thought to make the work safe.	• It is recognised that procedures cannot capture every potential eventuality.

References

Hopkins, Andrew (2010) Risk Management and Rule Compliance Decision Making in Hazardous Industries. Working Paper 72, National Research Centre for OSH Regulation.

Long, Robert and Long, Joshua (2012) *Risk Makes Sense*, Scotoma Press.

15 The miscellany

There are many other ways in which companies delude themselves that their safety performance is good. Most of the perceived positives are superficial because safety tools and practices are being applied without first developing the mind-set, the philosophy and the culture that is required to sustain great safety performance in the long run. They are not being safe, they are doing safety, and there is a substantial difference between the two. Alan Quilley, a Canadian safety consultant, calls this Appearance Based Safety, which sums it up perfectly.

Perhaps they paid a consultant to develop a nice, shiny binder full of safety management systems. Or possibly they have simply established some routines that keep the regulator at bay. It's also possible that they have been poorly advised and genuinely think that they are doing the right things.

I was once reviewing safety information submitted by a contractor to pre-qualify them to work on a project we were undertaking. I had a few emails, a completed questionnaire and a copy of the safety management system manual. The manual was great. It had everything that was needed and was well put together. Normally, I would have been quite happy with what had been presented, but something didn't quite feel right. There was some incongruence between what was incorporated into the manual and what was written elsewhere. So I looked a little more carefully through it again until I found what I was looking for. There was a cut-and-paste error in the manual. Buried deep within the text was a reference to another company that had no relationship with the one under review. The manual had been written by a consultant for a different organisation and copied across with the names changed. It had not been designed for the specific areas of concern of this business, took no account of its values or its culture and had not been implemented in any meaningful way. An off-the-shelf, one-size-fits-all nothing of a safety manual, of which both the company management and the consultant who sold it should be ashamed. Appearance Based Safety.

Here are a few examples that illustrate companies that think they are doing well, but in reality have some way to go. There are many others.

We have a safety policy

The safety policy is intended to be a high-level vision of the company's approach to safety. It sets out the board's (or owner's) intention for safety and states what steps they will take to achieve it.

There are many and varied companies operating across a wide range of industries and locations. Even when they have superficial similarities in where or how they operate, they all have wildly different cultures and approaches to the way they manage every facet of their business. Yet almost every company's safety policy looks the same. They are so banal in their approach as to be completely useless, telling little about what safety genuinely means to them. In turn this makes them so far from memorable that nobody can tell you what is in them, beyond some vague exhortation to be safe because we care. I once reviewed pre-qualification information for a government ministry for the 150 or so companies that were to be on their panel of providers. From those 150 companies' policies combined, there were no more than a dozen phrases repeated over and over.

The one item that sums up this pointlessness most effectively is the statement of legal compliance that you will find almost without exception in all safety policies. This is partly because it is actually a requirement of certain international standards for safety management systems. It consists of some form of the following, stunningly obvious statement:

> We will comply with all our legal obligations for safety.

Well, thanks for that. Here was I, as an employee, set to break dozens of laws because I didn't realise that our company policy was to be legally compliant. Thank heavens you told me. That could have been awkward. Now I'll have to limit myself to breaking all the laws in all of the other areas of our business where the compliance requirement is less categorically stated.

Really, what value does a statement add that simply states we're not going to break the law? I would imagine it's fairly implicit and doesn't need stating. If it does, perhaps that's not an organisation we should be working for. In some respects it is good that it is buried within a document whose tedious nature discourages any thinking, because it hardly sends a message that your safety is important to us because we care. It states that we do this because we have to – even if that's not the case.

If you must include a compliance statement to align with a badly written standard (and there may be some business imperatives of dubious value that require it), how about this?

> We go beyond compliance towards best practice to keep our people safe.

This implies the legislation is followed, thereby ticking that particular box, but goes on to point out that actually we care more than just to stop there. It's only a small change, but it helps to drive the culture we want.

I recently reviewed a guidance note issued by a regulator for the management of major hazard facilities (those with the potential for major explosions with catastrophic consequences). It stated that a good policy can be of significant impact in ensuring good safety performance. This is true. It then went on to list a large number of items that a policy must include, as well as some strongly preferential items, in order to be acceptable to the regulator. The majority of policy statements are relatively short, often single-page documents describing high-level expectations around safety. Most are shorter than the list of items prescribed. What will be the impact of this? Every company operating one of these facilities will develop a policy that almost exactly aligns with the guidance note so that they comply with expectations. This will result in the situation that every facility in the country will be operating *under the same policy*. This has now served to completely undermine the valid point about the importance of policy. How can a policy be of use when it is dictated by an outside organisation and takes little account of the specifics of the business, its operations, its culture, its intent and its direction?

I have no doubt this was not the intention of the writers of the guidance note but, too often, this type of guidance is provided by civil service policy writers who have little or no experience of how the corporate world will implement it.

A policy should reflect the value and the culture of the business, underpinning its safety direction so that everybody is clear on what safety means in their own context.

Go and look at your safety policy. Is there anything in there, other than your corporate logo, that links what it says to who you are and what you do as a business? If not, change it. Make it real and make it meaningful.

Safety is the first item on the agenda at every meeting

This is intended to show that safety is embedded into the organisation and, not only do we always consider it, we think about it first because it's the most important.

However, in many cases this will turn into getting safety out of the way quickly, so that we can concentrate on what we're really interested in. In any event, the reality is often that the last item on the agenda is most likely to be remembered and thought about after the meeting.

In an organisation where safety is a clear and constant focus, there is no need to force participation. Each item under discussion in a meeting is considered on its merits. If there is a safety concern or consideration, it is covered naturally. If there are no real areas of safety concern, the matter can be approached in isolation. Forcing the issue can be counter-productive and often results in 'safety moments' that have no real relevance and become another reason to disengage. If you have routine safety moments at the start of meetings, review them for a few meetings and consider how often they actually provoke interested and value-adding conversations around the topic.

I have worked in a business with a genuinely positive and good culture and safety focus, where safety moments were used in all meetings. Even there, I don't think more than 20 per cent of them had any follow-up discussion. If you are going to do this, why not re-brand them as 'safety conversations' and put the emphasis on asking rather than telling to elicit a response. Not only does this get the people in the room more involved, but it also forces the presenter to develop a topic that is more relevant to prevent it from falling flat.

Psychology research shows that repeated behaviours can drive attitudes and emotions, which runs somewhat counter to the intuitive idea that the way we think and feel drives how we behave. For most people now it feels wrong to be in a car without a seatbelt on. This is largely because of the forced compliance introduced in most countries in the 1970s and 1980s. Thus, routinely forcing participation can have a beneficial effect on involvement in safety over time. This needs to be carefully considered, however, so as not to go too far and turn people off, and should be viewed as an early-stage tool to help in the development of safety from a low baseline. It is not an indicator of good safety performance.

When someone proudly states that safety is the first item on every agenda, it is showing that they have some way to go in the journey to great safety performance. Whether it is a good starting point, or someone is doing it because they feel they have to in order to look the part, requires further investigation into the ideas and values behind it.

We have a blame-free culture

Blame-free cultures are intended to allow people to feel at ease in raising incidents. There is no hiding for fear of punishment, so incident reporting is strong and we can have confidence that all our incidents are being captured to allow investigation.

This is positive and certainly better than a blame culture, where incidents are not reported if at all possible, but it is merely an indication of an early phase in the maturity journey.

In a blame-free environment, where is accountability for our actions? We can quite happily disregard procedures and instructions, override or bypass safety equipment without fear of reprisals. Most people don't, of course, but a small minority do work unsafely despite appropriate training, equipment, leadership, supervision and work processes. If there is no consequence for their actions, what message does that send to the vast majority of the workforce who are conscientiously following procedures, assessing and controlling hazards and generally doing the right thing? Those people will feel disengaged and frustrated. Failure to deal with under-performance routinely appears as a significant concern in feedback from personnel. The inequity of doing a good job for no more benefit than the person who is not is a major source of dissatisfaction. People must, therefore, be held to account for their actions.

There will also be instances of unsafe work, however, where the individual was not at fault. Perhaps they have not been adequately trained in the task, were inadequately supervised or were inappropriately put under pressure to complete work quickly and took a safety shortcut as a result. Or possibly they made a simple and honest mistake, or forgot something inadvertently. In such cases, it may be harsh and inappropriate to hold them to account.

So what to do? Much work has been done following Reason (1998) in developing so-called 'just cultures'. As for many of the concepts in this book, this is widely known and understood in many safety circles, but is not necessarily practised in industry. A just culture requires effort in investigation to establish the root causes for incidents, so that it is clear whether the individuals involved were at fault and, if so, the degree of their culpability. This will range from totally blame-free victims of circumstance to wilful violation of safety protocols in full knowledge of the expectations and the potential consequences. The action taken can then be commensurate with the 'crime'.

Anyone who states the blame-free position for their business needs to understand that it is not a mature position and there is still a considerable way to go.

We investigate all our incidents

Incidents should be investigated. There is no argument about this. The principal reason for doing so is to stop the event, or a similar one, happening again. It is not to lay blame, but rather to understand the causes, so that appropriate action can be taken to improve. There may be times, as discussed earlier within a just culture, where one of the findings is that somebody acted inappropriately and action needs to be taken, but this should not be the key driver of the investigation. It is also very rare to find genuine malice or intent.

In many countries, legislation demands accident investigation. The completion of investigations is, therefore, simply minimum compliance, not an indicator of best practice. But completion of *good* investigations is.

Many companies have a short-form investigation report – a single page that captures the basic requirements for minor accidents. But for some of them, that is all they have. There is nothing more detailed for more significant accidents, so no detailed analysis is carried out.

Even for those minor ones, some overview trending is required to ensure that any similarities between incidents are captured and can be dealt with.

Most accidents occur through a unique set of circumstances. For minor or relatively frequent injuries, there may be some obvious similar elements, but there are generally few things that are likely to be repeated exactly. To fully understand accidents and uncover the potentially systemic faults that may lead to other incidents, it is necessary to undertake detailed root cause analysis. Simple investigation forms do not achieve this. At best, they may contain a checklist of 6–10 root causes to select from. But these tend to be assigned based on the judgement of the investigator, rather than any detailed formal assessment process.

Even when root cause analysis is done – and many of the larger organisations do have systems in place – there tends to be little rigour to the overall process. The root cause steps are followed according to the system that is being used, but little thought is given to selection of the correct investigation team, assessment and tracking of the quality of the output, or the actions identified, reflection on the effectiveness of the actions after implementation and so on. A lot of time and effort is spent in finding problems, with little spent on resolving them effectively. Worse still, some people seem to be fixated on the idea that there should be a single root cause, so they stop investigating when they find an appropriate culprit. Usually one that fits their preconceived idea.

'We investigate all our incidents' is a hollow statement, if the lessons from them are not adequately implemented or the process does not truly provide the depth of understanding required.

There is also a further question to consider for investigations. Many of the lessons learned from accidents can also be learned from near-miss events where no injury occurred. There tend to be more of these – some extremely serious. Are your near misses being adequately reported and investigated? This can be taken a step further to look at events occurring during normal operations where seemingly no 'incident' took place. We will touch upon this later.

We are certified against a standard

The use of national and international standards to demonstrate how good we are has become increasingly prevalent over the last 20 years or so. In many areas, particularly government and local authority contracts, it is not possible to make the tender list without some form of certification. Many service providers and manufacturers proudly display logos of certification as badges of honour.

Standards originally arose out of engineering and manufacturing, where it is important that materials and methods of design, production, construction and installation are robust and repeatable. More recently, though, led by the ISO 9000 systems in quality management, they have been adopted into management systems where there are two fundamental problems with certification by standards.

Problem one is the obvious one. Every organisation is unique. Its products or services are unique. Its culture and people are unique. Therefore its management system should be unique. To suggest that the management system for an airline should be the same as a spring manufacturer or a consultant is absurd. For this reason, the management system standard provides a framework of very basic requirements only, as it must do to be applied across different industries. The standard can provide a reasonable starting point if building a new system, but beyond that it provides little guidance or help.

It is possible for a management system standard to be an effective and useful tool. The Australian Standard for the management of oil and gas pipelines, for example (AS2885), incorporates a well-defined management system approach that is rigorous and robust. Compliance with it will go a long way to achieving what is required for a well-run system. But this is for a single industry in a single country, and it incorporates some aspects that are more difficult to apply in some countries other than Australia. This level of rigour is simply not possible for a standard that applies across many industries and locations.

Problem two is the capability and independence of the auditor. A good auditor is well worth the investment. Having an independent review of processes can be very valuable. But many certifying auditors are only interested in making sure that all of the systems line up with the standard. Having a system in place does not mean that it is being properly implemented, and audits tend not to be of sufficient length or depth to fully investigate this. Similarly, not all auditors have the capability to go beyond reviewing documents to make sure that procedures are in place. An independent audit is good, but will only find worthwhile improvements if the company being audited is open to that finding being exposed. If certification is necessary to bid for contracts, it is potentially damaging to expose flaws as these may result in certificates being revoked. The audit then becomes a process of presenting the best possible side of the business to the auditor, rather than exposing areas of weakness to identify improvements. This is the opposite of its intent and is certainly not a genuine tool for learning and improvement.

Furthermore, certifying companies are in competition with each other for the business. If one auditor is finding concerns that may lead to non-certification, it can be easier to switch to a different auditor than it is to fix the problems.

Of course, good organisations welcome the auditor and welcome findings for improvement. But those types of organisations already have the culture, leadership and systems that mean certification is unlikely to be revoked anyway. Companies fall simplistically into two categories – good and bad. Over time, the market will sort out the bad ones. The certification process turns that into three categories:

1   good organisations who would audit themselves irrespective of the existence of certification because it is of benefit to them in the longer term;
2   poor organisations who manage to pass the audit through providing an unrepresentative impression of their capability and therefore look good even when they're not; and
3   organisations that are so bad, they can't even pass a baseline audit.

This three-category position seems worse than the two we started with. The market will still sort out the bad ones, but some of them will get more traction than they otherwise would have.

Note that this is not to denigrate the importance of a good management system. In fact, this is one of the most important aspects of developing strong safety performance, as we shall see later. But management systems being standardised across a range of companies is not a best-practice innovation. Conformance to a standard is much like compliance with the law – it is a baseline position. A good starting point only. Many an organisation has been fooled into thinking they have good performance by their external certification audit results.

We pre-qualify our contractors

A significant proportion of businesses execute work using contractors brought in to provide either manpower, specialist experience or both. To ensure that the contractor brings a suitable level of safety knowledge and performance, they are often put through a pre-qualification process. This usually consists of completion of a questionnaire together with provision of supporting evidence for the answers given. This is very similar in principle to the process of auditing against a standard as discussed above. Indeed, many organisations use certification in lieu of a specific pre-qualification process.

Pre-qualifying contractors is a sensible thing to do; however, in practice there are a number of shortfalls that are commonplace.

Setting the bar too low

Often, the level of pre-qualification required is so low as to be almost useless in discriminating between good and bad performers. Contractors are required to provide a copy of their safety policy, their accident statistics (almost impossible to verify) and fill in a short questionnaire that usually does not go beyond confirming that hazard identification and accident investigation processes are in place.

With such a low bar, almost nobody fails and so the process is almost completely pointless, wasting everybody's time and effort – particularly the contractor who has to fill in dozens of slightly different variations on the same form for different clients. If you have a pre-qualification process in place, investigate what proportion of suppliers fails to meet the target and see if it is adding any value.

Not being risk or task based

The normal method is for the contractor, having successfully jumped over a very low hurdle, to take their place on a supplier list of approved contractors. This is usually held by the procurement department and, again, is sensible in principle. Unfortunately, that list almost never includes any specifics as to the particular activity the contractor was bidding for when undertaking the pre-qualification. What happens in practice is that when someone requires a contractor, they choose one, check that they are on the list and then begin contract negotiations.

But the level of safety performance and capability required is hugely dependent upon the task at hand. If the job entails dealing with asbestos, it is important to know that the contractor has that experience. Typically, there is just a yes/no against a low baseline with no specifics about the level of risk or specialism required.

The pre-qualification process should vary according to the risk and the task being undertaken. An electrical maintenance contractor should face a higher level of pre-qualification than the contractor who comes in to water the plants. This also prevents the 'one size fits all' approach that many smaller contractors, in particular, complain about. The supplier list should then include a clear statement as to what level and type of task each contractor is approved for.

No feedback loop

As the process is typically a desktop document-based approach, it really gives little information as to the actual capability, as this is so dependent on the people on the ground undertaking the work. This can (and usually does) bear little resemblance to the paperwork. At the end of each project, or contract, each company's performance should be reviewed and the findings added to the supplier list. This allows performance to be trended over time and provides warnings or reassurances for future jobs. In this way, there is a genuine incentive to perform and to improve over time.

Overall, the general approach to pre-qualification is that it provides a proxy for capability and is, therefore, unfortunately too often used as an unspoken justification for not carrying out audits, reviews or establishing standards with contractors. But the process is usually nowhere near robust enough to be anything other than a convenient excuse for abrogation of responsibility towards contractors. The majority of businesses treat contracted work as something they can throw over the fence and forget about while it is carried out by second-class citizens. Nowhere is this attitude more clearly expressed than when companies report on accident statistics that don't include work undertaken by their contractors. On occasion this is appropriate, but most of the time it is a failure to acknowledge the client's role in specifying work and driving behaviour in their contractors.

Challenge yourself

With any of the claims you make in relation to the activities you do and the approaches that you take, challenge yourself as to their effectiveness. How directly and immediately does what you do impact upon the actual management of risks in the field? Does investigating your incidents decrease risk? No. Does investigating your incidents well, clearly identifying causes and

implementing effective corrective actions decrease risk? Yes. Does telling a
room full of disengaged people during a safety moment how you wear safety
gear while doing DIY at home help to reduce risk? No. Does having a good
and engaging safety conversation with the same people about a specific issue
that has direct relevance and impact on their work area help to reduce
risk? Yes.

Key points

From this...

- Safety activities are used to con-
  vince the business that safety is
  being done well.
- Audits and certifications demonstrate performance.

...to this

- Safety activities are routinely
  reviewed and challenged for their
  effectiveness in reducing risk.
- In-depth and quality application of
  value-adding processes demonstrate performance.

Reference

Reason, James (1998) Achieving a Safe Culture: Theory and Practice, *Work and Stress*
    Vol. 3: 293–306.

PART II SUMMARY

- Many activities that we do in the name of safety are implemented out of habit, rather than through a careful consideration of their role in reducing risk.
- There is an over-reliance on systems and procedures and insufficient acknowledgement of the role and skill of the workforce in carrying out work safely.
- Systems and activities are implemented by viewing them through the lens of management, rather than spending time to understand the impact on the workers and how they will be perceived and received.
- Company management needs to become more familiar with, and better versed in, the detailed aspects of safety theory.
- Safety managers must become more rounded in their understanding of, and contribution to, the business as a whole.
- Stop managing safety by strap-lines, mottos, numbers and posters, and start actually understanding the day-to-day issues of the workplace.

In Part III, we will look at processes to follow to implement some better approaches to safety, taking into account the local context of a particular organisation.

Part III

Your context

16 The bespoke approach

So, having established some of the things we should and should not be doing, how do we achieve the performance we want in our organisations? Safety should not be viewed in isolation from other functions and for successful implementation you need to take a holistic approach to your systems and your leadership.

There are some key considerations – some of which we have already discussed and others which are to follow in more detail:

- Excellence is a habit, not an act. It is not a pick-and-choose philosophy. You cannot choose to be excellent at safety, while ignoring poor behaviour in other functional areas. There cannot be a stand-alone safety culture divorced from the general company culture. It is all or nothing.
- Every organisation is different. While there are key factors that are constant, your processes, systems and approach should be bespoke, tailored to your requirements. By all means bring in a consultant or subject matter expert to help in your development, but this must be your system rooted in your culture, developed in conjunction with your workforce and genuinely believed in by your leadership.
- Do not over-develop the system. Remember that keeping the workforce engaged is the single biggest factor in your success. Systems should be put in place for those items that are critical for output. Trust employee selection, competence and training to manage those areas that are core skills.
- The final system and approach should resonate in such a way that following it and working within it feels like the right thing to do for everybody.
- Excellence is not perfection. If you wait for perfection, you will never begin. Get the system up and running and then refine as you go. Do not be afraid to change things if what you established is not quite working correctly.

What follows is a guide to take you through the process of developing your systems. Not everything will be right for you. Some may be inappropriate. Some may be over the top for the level of risk you face, or for the scale of your organisation. Think critically about what you need and what these things offer and adjust, remove or add as required.

Because context is everything, this guide is a process to follow and some questions to ask so you can establish your own needs. It is not an answer. I'm afraid you're going to have to work that out for yourself. Sorry.

There are three key steps in developing and maintaining your approach:

1   establishing leadership;
2   setting the culture;
3   developing and implementing the system.

Note that while these are broadly sequential, it will not be as straightforward as simply stepping through and ticking each off in turn. There may be a number of iterative loops where circumstances dictate returning to an earlier phase and adjusting it slightly. Or some parts may run in parallel. These are all perfectly appropriate and, ideally, will be stressed at the beginning of the process so all those involved are aware that any such backtracking is about refinement as we learn the best approach, not about doing it again because it was wrong.

The most important thing is to ask good questions, have good conversations and use these to develop a cohesive, supportable, sustainable and genuine approach to improving safety.

Of course, this part is not as much fun as poking holes in existing practices and being generally critical; but it doesn't do to go around telling everyone what they shouldn't do without offering a viable alternative (good life lesson).

So, enough explanation. Let's get into it.

17 The leadership bus

In his book *Good to Great*, Jim Collins (2001) stresses the need to bring good leadership into a business in order to develop it into the most successful organisation that it can be. The emphasis is placed on finding the right people and getting them on board. The role they will take within the business is secondary; the most important factor is that they bring the leadership and culture that will work for the organisation. Clearly, there is the need to balance this with the right technical skills; the more specialist or complex the industry in which you operate, the more this balance will swing towards the technical component. As for safety, in business management there are too many grey areas for an absolute rule to apply. You can have as many fantastic leaders in place as you like, but if none of them understand the marketplace or the technology, the business won't be successful.

Nevertheless, the guiding principle still applies. If there are two well-qualified candidates for a senior role and one has a stronger technical bias, while the other has stronger leadership skills and a better alignment with your intended culture, go for the second.

It is vital that the CEO takes the lead in this development process. Without the right leadership in place from the very top, it will not be sustainable. The CEO must understand what their requirement for excellence is and how it should manifest itself across the business and ensure that the leadership team is fully on board. If they are not ready or willing to take this journey, then get them off the bus and bring in (or promote) someone who is.

At this stage, this is not a safety process. It is simply a leadership process. Remember that safety culture cannot exist in isolation from the broader business culture, so the same foundational principles will apply.

Once you are comfortable the right people are around the table, you can begin the safety conversation.

Safety, as we have seen, is difficult. It is a wicked problem – that is, one without an easily definable solution (if any). It is an emergent property of a complex system. In short, it's tricky and you can't solve it by thinking in straight lines. You need to come at it from multiple directions to catch it off guard and give yourself a fighting chance. To investigate it fully and to place it properly in all its complicated context, the best approach to start is to have a conversation. Conversations are amazing. Try to produce a document or a presentation that captures all of the aspects and it will be almost impossible to follow. But develop it as a conversation and the brain takes all the multiple strands, different viewpoints and contrasting opinions and places them coherently into a multi-dimensional picture that makes perfect sense.

This conversation must be completely honest. Any successful leadership conversation should be, of course, but there is a tendency in safety to get hijacked by the ethical and moral dimension and say what you think you ought to, rather than what you genuinely believe. This must be resisted. Differences in opinion can be accommodated, hidden opinions cannot. An independent facilitator can help in keeping the conversation moving and providing challenge in the right places. But they must be a strong facilitator. There will be a tendency to try to begin to solve the problem, to get into tools and activities. Many leaders have a bias towards action and feel uncomfortable leaving a session without a task list. The facilitator needs to be able to jump on that and drag the focus back up into culture and leadership, and in a C-suite setting there are typically a lot of powerful egos and smart people to corral. There are many 'how to have a successful meeting' guides that emphasise structure and control. They focus on agendas, actions and whether what you are about to say is directly applicable to developing a specific outcome for the topic at hand. This is okay for routine operational meetings. It is not the way to have a strategic, vision-setting meeting. Innovation, ideas and understanding do not come from rigidly controlled, agenda-driven meetings. The facilitator should make sure digressions have some value, but should not cut them off too soon.

The purpose of the conversation is for everyone to understand the context and to know what the overall vision for safety improvement is; but most importantly to agree on why we are doing this. However the messages and processes are developed and rolled out later in the process, they will be more easily, more engagingly and more authentically delivered if there is a clear and well understood *why?* behind them.

Simon Sinek, the leadership expert, talks about engagement coming from the *why*, rather than the *how* or the *what*. The message should always come from the *why* (watch his TED talk presentation).

Not only is this more powerful in terms of engagement, it has the added advantage of being more or less timeless. Our processes may change, our tools may be refined or replaced, but our reason for doing it will stay the same – or at least be very slow to change. Hopefully, in the development of a safety *why*, the conversation will eventually come around to the fact that, actually, we care about our people – our colleagues and friends – and we don't want them to be harmed in any substantive way. We may rephrase that or couch it in different terms over time, but it will never change. You won't wake up one morning and say, 'Actually, I do want my staff hurt.'

The conversation should not be rushed and should cover a range of topics, including the following.

What is our current reality?

How well are we currently performing? How do we, as a leadership team, influence that performance? Do we understand what our key risks are? Are we adequately focused on those? Do we have a formal safety strategy or are we being reactive? And so on.

This part of the conversation is simply a wide-ranging discussion about current issues. It can be largely self-directed by the team, but the facilitator should identify and highlight those points that may demonstrate biases in thinking. It is important within this part of the discussion that the leadership team reflects on its own role in establishing the current reality and makes the leap to hold itself accountable for making the first changes.

What are the current constraints on performance?

What is it that is stopping our current performance from achieving what we want? Who are the blockers? Do we recognise the fact that our performance is a reflection of our leadership? This can be a challenging discussion – especially if we've just decided that our current reality is a long way from desirable. This is why it is important to establish the current reality first, before anyone senses their own culpability in it and begins to dissemble. Because once it is acknowledged that we get the performance that our leadership deserves, it becomes obvious that if we want to change, we need to examine our own performance, behaviour and leadership activities first.

Having established that constraints are in place, we need to consider whether those constraints are real or imagined. Whether they are immovable or can be bypassed. There are no solutions at this stage, but we may identify which are the most significant and highest priority for dealing with.

How do we define safety?

It always takes a little while for strategic conversations to warm up. People need to pull their minds out of the operational day to day and readjust to broader thinking and longer time horizons. Make sure the conversation is well established before you throw this one in there because it is a little bit philosophical.

As noted in Chapter 4, we all bandy around the word 'safety' at work (and at home), but don't stop to examine whether we are all on the same page in doing so. Is my version of safety the same as yours? How safe is safe?

There may be some resistance to this part of the discussion ('it's just semantics' I hear them cry), but it is important to have the debate. Not only to recognise the different viewpoints on the bus, but also to anticipate the perspective of the workforce that will come on board later.

Typically the views will start with 'being free from harm' (for which see the zero paradox). This is usually short-circuited by pointing out that we can often act unsafely without being hurt. The conversation will then move to somewhat more considered positions relating to acceptability and acknow-ledging the grey areas of risk–benefit trade-offs. This becomes a very worth-while debate in the context of your business. Even if the views do not entirely converge, the debate is useful to exemplify the complexity. How can we keep people safe if we can't even agree on what that means? A little variety is okay (after all, some of us bungee jump and others would never take the risk). Too much variety needs to be resolved.

What is our desired future state?

This should naturally follow on from the previous discussion. Generally speaking, everybody wants their people to be safe. Having established what we think that means, it is a fairly small step to defining a desired future state. At this stage, there is a temptation to capture it in a catchy slogan. You will have seen them all and we have touched on some of them in Part II. Beware both the zero paradox and priority confusion in developing this.

I have a general difficulty with 'branding' safety too much. We don't seem to have similar branding in other functions, so I see no reason to separate safety out in this manner. However, it can be useful for some form of shorthand to be employed to help conversations throughout the workforce. For example, we could give our safety management framework a name – let's call it HomeSafe – so that rather than continually referring to 'the safety management framework', we could refer to 'HomeSafe'. But this must be a shorthand that is based on an in-depth understanding of all that it represents, not simply a slogan. People will then interpret it within the context of a clearly understood broader picture, so that when it is employed it conjures up appropriate thinking that deals with all the aspects required, rather than reverting to lip service.

Whether or not a slogan is developed (and it may not be at this stage anyway), this part of the discussion should be the easiest.

How will that manifest itself day to day for the leadership team and the wider organisation?

It's easy enough to make general 'motherhood' statements about how our safety performance will look, but what does 'being the industry leaders in safety' or similar actually look like in practice? What will the leadership team and the workforce experience, see and hear that demonstrates that the future state has been reached.

This is where your understanding of leading indicators (see Chapter 9) begins to bear fruit. We have seen that it does not really tell us anything about your safety performance to focus on injury statistics and lagging indic-ators. Good safety performance will manifest itself in leading indicators, in positive attributes and in nuggets of good leadership. Note that not all of these will be easily measurable in the traditional sense.

Such indications may be simply that workers feel comfortable raising safety issues; there may be engaged and effective safety committees; safety interventions are viewed as positive and helpful; audit performance is good, with audits viewed as helpful learning exercises; formal investigations may occur into projects where performance has been particularly good to understand why and how; and so on. All of these are symptomatic of a positive culture and an engaged workforce.

What are the key steps in getting from our current reality to our future state?

Without getting into too much detail at this stage, understand the key, high-level steps that are required to make it to the future state. These may include:

- redefining the safety leadership approach within the organisation;
- establishing a clear safety strategy;
- setting up a safety improvement working group;
- investigating new tools for safety management;
- hiring a new safety manager;
- refreshing workforce engagement and feedback mechanisms;
- properly defining and understanding the highest risks in the business;
- and so on.

By the time the conversation is complete – and it may run over more than one session, depending on the available time in diaries of busy executives – the whole leadership team should be aligned, engaged and understanding of the direction to be taken.

This exercise is crucial and should not be shortcut or bypassed. If it is, blockers will appear later and derail the improvement process. This may be in the form of budgetary cuts, lack of availability of key people or sections of the workforce that are disengaged and not supporting or participating. Good safety crosses all organisational boundaries and all parts of the business. Every area needs to be on board and travelling in the same direction for it to be successful.

Key points

- Ensure the whole leadership team is fully and honestly engaged in the safety improvement process.
- The leadership team needs to own the process and take responsibility for where the business is now.
- Clearly define the safety vision for the organisation.

Reference

Collins, Jim (2001) *Good to Great*, William Collins.

18 The leadership attribution

We have talked at length about the importance of leadership within safety. This is, of course, no different to leadership in any other part of the business. Without good leadership, you're not going anywhere. But what does it mean? In the previous chapter we talked about engaging the leadership team to make sure we're all going in the same direction, but how do we know we have the right people on board?

Anyone with any interest in learning about leadership will have seen and read numerous articles about the difference between leadership and management – leadership being about setting vision, expectation and culture, while management is about organising, monitoring and resourcing. Both are important, but serve different purposes. Typically (although by no means necessarily), there is more leadership the higher the position in the organisational structure.

In safety, we often talk about 'leadership at all levels' as there are people anywhere within that structure that can provide leadership. I have some general misgivings about this. Not as a philosophy, but as a practical reality. It falls under the same umbrella as 'empowerment' and the oft-quoted 'safety is everyone's responsibility'. Unfortunately, while noble sentiments, these more often than not turn into an opportunity to blame someone when things go wrong. Workers are asked to show leadership, but are not given resources, authority or autonomy to do anything about issues that concern them. They are empowered, but not usually with any clear definition or guidance as to how extensive that empowerment is; when decisions made will be supported or when they will be shot down.

For the purposes of this discussion, then, when we are talking about leadership, we are still talking predominantly about senior leadership from an organisational perspective. The attributes can be adopted or exemplified by anyone at any level, but the focus is towards those who are structurally in a position to influence.

As we have seen, safety crosses into many different areas of specialism that warrant entire libraries in their own right. Leadership is another of these and a short chapter here is (obviously) not intended to cover all aspects of leadership theory and practice.

We often hear, though, about the requirements of safety leadership. Typically these include a range of aspects such as:

- visibility in the workplace;
- setting clear objectives;
- putting a policy in place;
- getting involved;
- being committed (whatever that means);
- ensuring budgets are available;
- measuring outputs.

But these are all very transactional activities. They are more management than leadership. True leadership is *transformational*, not transactional.

Leadership in safety is reflective of leadership in general. The attributes of a safety leader are simply those of a leader in its general sense. These are founded in the *why* we discussed in Chapter 17. The *what* and the *how* of safety is generally looked after within the management discipline rather than leadership.

There are dozens of books and studies available that are based on the review and assessment of successful leaders in order to determine those magic ingredients that can turn you into a world beater. There is a flaw in most of these in that they find who is successful and look for consistent factors, but do not identify and consider any control groups. In a correctly developed scientific review process, a control group is necessary to corroborate the findings of the study group. This is another manifestation of the correlation versus causation problem. Simply the fact that members of an elite group all exhibit a similar characteristic is no demonstration that it caused their membership of that elite group. For example, it is often quoted that 'successful people are early risers'. This is usually supported only by anecdotal evidence that this or that CEO, world leader or superstar has held three board meetings, been to the gym and fashioned an abstract sculpture by 6 a.m., when the rest of us laggards are only thinking about getting up. The implication is that you should get up earlier if you want to be successful.

However, to understand this properly, a control group is needed of unsuccessful people (I'm not sure how the advert would be phrased when looking for volunteers). Does the early-risers-to-night-owls ratio differ between our successful and unsuccessful groups? Or we could find a group of all early risers and find out what proportion of them are 'successful' to understand the degree of causation. To conduct such a study and to further control for all the other myriad factors involved in success would be extremely complex and time-consuming (not to mention expensive). Far easier to review a group and look for similarities. In such a circumstance, the best we can hope for from a scientific and objective perspective is to say that if all the successful people do exhibit a similar trait, it is unlikely to be counter-indicative of success. So it's a good start. But qualitative reviews of 'great leaders' are not necessarily predictors of great leadership when employed in mimicry.

There are two counter-arguments to this. The first is that leadership is very subjective and is based on relationships and perspective as well as factual and measurable components of success. In which case, as long as it works for you and provides you with the environment and inspiration to perform to your best, then it is good leadership. In that context, anecdotal evidence and personal interpretations are, arguably, as good an indicator as anything else.

The second is that, particularly for a large data set, consistent and repeated core components of findings can, in themselves, provide adequate information to interpret. Especially in dismissing those aspects that do not contribute to success. For example, a review of all Olympic gymnasts shows that none are overweight. This does not mean that if you lose weight you too could become an Olympic gymnast (although TV shopping channels may have you believe otherwise), but it is sufficient evidence that if you are overweight, you definitely could not. In leadership studies, there has been enough volume and sufficient meta-analysis that while we cannot definitively say that certain attributes will guarantee good leadership, we can certainly point to aspects that clearly increase the likelihood.

There is therefore enough evidence of core components of good safety leadership to conclude that, without them, leaders will not achieve what they could.

Enough pontificating about the quality of the data. Here are my six key attributes of a safety leader based on my own experience working with (and within) a wide range of organisations and a significant number of leaders, cross-referenced with reviews of numerous leadership treatises and theories.

Effective safety leaders:

1   understand and acknowledge their role in performance;
2   understand the interconnections between safety and business performance;
3   are transformational;
4   are learning-oriented;
5   understand the trajectory of their messages; and
6   are authentic.

There are six, which happens to be the number of letters in 'safety' and also 'leader'. It would be nice and memorable (not to mention marketable) to turn the list into an acrostic based on the letters in the word, but life's not like that and there are already too many examples of over-simplification branding exercises forcing names into neat packaging.

Understand and acknowledge their role

We discussed this briefly in Chapter 17. Most senior managers recognise the importance of their leadership input into future success, even if they may not necessarily be able to articulate or practise how best to achieve that. Fewer are prepared to acknowledge their role in *how they got here in the first place*.

There is an immediate tendency to discuss how everyone else needs to change. It takes a great deal of maturity to look in the mirror and start the change process internally. This is true of much change – not just in safety.

But, quite clearly, logic dictates that the change must start within the leadership team. If our success in two to three years' time is largely dictated by the strength and quality of our leadership over the intervening period, then it follows that our current position must have been dictated by our leadership over the foregoing two to three years.

Effective leaders understand and acknowledge their role and are prepared to lead from the front by changing themselves before changing the organisation. The natural state of the worker should be to be engaged with safety – after all, they do not wish to get badly hurt or killed at work. The safety leader realises this and changes what they do that is currently disengaging them.

The connections between safety and business performance

There are any number of reports available that correlate good safety perform-ance with good business performance. We discussed this earlier in Chapter 7. The best safety leaders recognise, acknowledge and act upon the connection between safety performance and business performance. It is worth repeating, because the fundamental principle underlying this inter-connectedness across the business is crucial. If you take nothing else from this book, remember that safety culture does not exist in isolation from the broader business culture:

> The principles, practices, disciplines and culture that underpin good per-formance in safety are the same as those that underpin good performance in quality, maintenance, customer service, business delivery, staff engage-ment and all aspects of your business. If you want to improve safety, first improve those fundamentals across the whole organisation.

Effective safety leaders understand this and work on the business as a whole to get improvements across the whole enterprise. They also recognise that, in the course of working on this across the business, safety is a really good place to start. So much of the foundation of good culture is based on trust and caring for the team. A genuine and effective improvement in safety provides an exceptional platform on which to build.

Transformational

Great leaders are transformational rather than transactional. They use vision and empowerment to inspire teams to take ownership of issues and deal with them themselves, encompassing and bringing out all the talent of the organisation.

This is not to be confused with the high-energy, motivational, charismatic manager who is all talk. Inspiration and motivation fall over quickly if they are not backed up and supported by resources, ongoing encouragement and, importantly, acceptance of positive failure (i.e. those failures that occur necessarily at some point in the pursuit of improvement and responsible risk taking).

Transformational leaders are people-centric and collegiate. They are prepared to listen to advice; encourage their teams to grow and develop; share credit for outcomes and are consultative and cooperative, although not afraid to make a call on difficult decisions. It is a difficult balance to strike and, therefore, rare to find.

Daniel Pink (2009) has reviewed motivational theory and practice (another excellent TED talk as well). He states that traditional incentive methods designed to motivate are not only ineffective, but can actually reduce performance. Experiments have shown that external motivators, such as financial bonuses for completion rates, work well for simple, routine, mechanical tasks that require little or no thinking or creativity. For more complex tasks, a bonus for completion actually reduces the rate at which problems are solved. The reward focuses the mind on speed and limits it in the imaginative and lateral thinking required to find a solution. In today's world, difficult problems requiring creative solutions are far more common and are typically at the heart of genuine improvement processes. This is certainly the case in safety.

So, if bonuses do not work, what does? Pink highlights three areas that motivate:

- autonomy
- mastery
- purpose.

Consider these in the light of the description above of a transformational leader. Autonomy – allows people to deliver their own solutions and take appropriate risks. Mastery – encourages the team to grow and develop. Purpose – establishes a clear and compelling vision that inspires the team to perform.

Contrast this motivational theory to the majority of safety management activities and attitudes prevalent in the workplace – follow the procedures: don't break the (many) rules; investigate and blame when accidents happen; only talk to the workers when something has gone wrong; 90 per cent of accidents are caused by you making a mistake. Is it any wonder that people are demotivated and turned off to safety?

Learning-oriented

As noted above, one of the key motivational characteristics is mastery. Effective leaders take the time and spend the effort to learn more and understand more, therefore, increasing their own mastery.

This manifests itself in a questioning attitude, in lifelong learning, in seeking to understand difficult problems.

One of the problems in safety we have discussed is a tendency to attempt to over-simplify and not acknowledge the complexity of the workplace and the people within it. Effective leaders will simplify to aid understanding, but only as far as is realistic. One of the many quotes attributed to Albert Einstein is that, 'things should be made as simple as possible, but not simpler'. I don't know whether he actually said that, because if he said everything that is attributed to him on the internet, he would certainly never have had time to develop his relativity theories. But it suits this discussion. It is also often said that the best way to genuinely understand something is to explain it to someone else. To understand, simplify and explain well requires mastery of a subject.

An effective leader constantly searches for more knowledge and understanding. They do not take matters at face value – they explore and question and challenge. This includes being open to learning from elsewhere within the organisation – listening to their workers and taking advice from those with a better knowledge of the practicalities of the workplace.

They also show the organisation that they value learning. They support training; they encourage mentoring programmes and shared learning events. This has a cascading value throughout the business. When the whole business is adding to their knowledge and understanding on a routine basis, the improvements it can bring are dramatic.

Understand message trajectory

This is a more subtle attribute than the others and requires some context.

When a leader sends a message, the business responds. How, when and where the response occurs is the trajectory of the message. There are three key factors within this. First, not everybody responds in the same way. Second, as a senior leader everything you say and do and the behaviours that you model are scrutinised and interpreted by the business. This also includes things that you don't say or do. Nature may abhor a vacuum, but not as much as business does, and people will fill spaces with rumour, conjecture and sometimes downright nonsense if gaps occur at important junctures. Third, as a senior leader the feedback you receive from the business is not the truth. The better your culture and openness, the closer to the truth it will be, but information is simply not passed back up the hierarchy without at least some sanitisation.

An effective leader recognises that different parts of the business will respond in different ways. Sometimes, this can simply be acknowledged – vive la différence – and we can move on. At other times what is helpful and positive for some may be damaging and insulting for others. In safety, there are typically wide variations in capability and experience within different business units. In high-hazard industries those in the field are highly experienced, risk aware and technically and practically competent, routinely facing and effectively managing significant risks. They require a very different message and approach to office-based workers who have a much lower risk profile. In contrast, when considering work-related stress due to long hours, fatigue and pressure to deliver, these risk profiles can reverse in certain environments.

I once worked with a business that sought to deal with both stress and general safety awareness through a series of whimsical cartoon characters looking surprised when wearing high heels instead of safety boots, or taking a lunchtime walk to manage their stress. This had the result of patronising and insulting those in the field by belittling the risks they faced and insulting their mastery. At the same time, the over-worked and stressed-out office staff were upset that there was no recognition of the causes of stress, putting all the onus of dealing with it on the workers. In both cases, the response of the workers was entirely predictable, had they simply been thought about in terms of the message trajectory. The message could have been amended and targeted to respective groups to appeal to their needs.

Messages are not limited to deliberate awareness campaigns and emails sent to staff. The symbols we use, the actions we take and the decisions we make are all interpreted by the business. Semiotics is the study of signs and symbols and their impact on communication. This should not be forgotten about as the impact can be very powerful. Signs and symbols promoting safety compliance completely undermine a spoken message of autonomy and empowerment. As we touched on in the Priority Confusion, when you send a message, it is crucial that it must be one that can be supported in the long term by your actions. All leaders know to 'walk the talk' but not many consider what the 'walk' will look like when they are designing the 'talk'. This is the message trajectory. The question becomes: 'If I say, do or show this message, what expectations of my actions does it set up in light of how it will be interpreted?' If the message is poorly designed, the leader either has to act inappropriately in order to remain congruent, or be inconsistent.

Finally, when feedback is received on messages, read between the lines. There is a classic meme on communication that has been doing the rounds since before they were called memes. This shows in a humorous way how a message is misinterpreted moving up through the communication channels, but it is surprisingly accurate. I have included it as Appendix 3.

Treat all feedback with care, taking into account the source and the messaging context. I once saw a blog on a company intranet asking what could be done to win over the 20 per cent of the staff that did not believe zero harm was possible at work following a survey. Based on the message trajectory, this was the wrong question. Zero harm was promulgated zealously within the business, was part of its core values and was evangelically promoted at every turn. In such an environment, the real question should be: 'If 20 per cent of people are prepared to say they disagree with it, what is the real number of people who disagree but recognise it is not in their best interests to admit to it? And what does that number say about zero harm as a core value for us?'

There are many complications and unforeseeable outcomes in trying to understand the trajectory before sending the message. It will never be perfect, but the effective safety leader avoids the most significant issues and deals with any other fallout through their established openness and collegiality, withdrawing something (with explanation) if it turns out it was the wrong approach to take. For examples of when this has failed, see the myriad of consumer/public outrage Twitter storms prompted by companies that have failed to think their message through.

Authenticity

And lastly, effective leaders are authentic. Their concern for their people is genuine. While recognising commercial realities, they do not need to see a cost–benefit analysis for every safety initiative. Sometimes the benefits are intangible and that's okay, because our people matter and we do this because we genuinely care.

There is no need to labour this particular point. Lack of authenticity will usually reveal itself before too long and this, more than anything else, will stop improvement progress in its tracks.

Key points

- The leadership team needs to work on themselves before they work on the business.
- The underlying principles that improve safety work across all functions – improve the whole business rather than safety in isolation.
- Set a clear and compelling vision for safety excellence and give people the space to deliver on it.
- Think carefully about messaging before it is delivered.

Reference

Pink, Daniel (2009) *Drive: The Surprising Truth About What Motivates Us*, Riverhead Books.

19 The culture cascade

Now that the leadership team is fully aligned, we need to engage the rest of the business. The purpose here is to develop the semi-mythical state that is the business culture. One of the reasons the leadership alignment is so key to this is that culture is heavily influenced from the top of an organisation. The first discussion has hopefully removed any potential cultural blockers. If a member of the leadership team is not aligned it can isolate the entire section of the business for which they are responsible.

Much is written about corporate culture (and safety culture) and what it means, but at its heart, cascading culture through an organisation can be distilled into two short phases:

- What interests my boss fascinates me.
- Monkey see, monkey do.

It makes little difference what is written in your company's stated values or list of desired corporate behaviours. The first part of what drives the culture are those things that the leadership asks about and shows genuine (not lip service) interest in. The desired culture can then be embedded by showing interests in those areas of performance that are part of that culture. The second aspect is the behaviours modelled by the leadership. What employees see their bosses doing is what they will replicate – monkey see, monkey do. In combination, these develop the top-down part of the culture. This is of course a gross over-simplification, but it serves the purpose of this discussion.

Top down

A consistent approach by an aligned leadership team sets a cultural tone that cascades to the next tier of management who, fascinated and copying, pass it to their teams and so on through the structure (Figure 19.1).

This process can become unstuck when cultural blockers appear. Cultural blockers are individuals within an organisation who have a different view to the mainstream culture and are sufficiently influential to model different behaviours that impact on their part of the business. Note that this influence

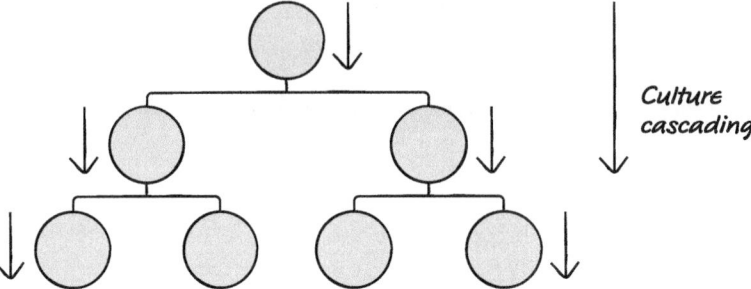

Figure 19.1 Culture cascading.

may not be formal due to their role, but may be due to their nature and character. There are many individuals within organisations that influence without authority – ironically a strong leadership trait if properly harnessed. If their degree of influence is stronger than that of their manager, then the culture cascade will be blocked in that area (Figure 19.2).

In this instance, the best solution is to involve that influencer and bring them on board the bus. If they have strong influencing skills that can be brought to bear in the right way, they will be a valuable asset to the business. If you know your business well, you should already know who these informal influencers are and bring them into the cultural development process early on. If they are part of the development process, they are far more likely to support its application.

Failing this, put stronger leadership in place in that part of the organisation who will be able to gain greater traction and overcome the blockage. As a last resort remove the blocker either to a different part of the organisation, or by offering them the chance to join a different company whose culture is a better fit (corporate euphemism alert!).

It can be clearly seen, then, that for developing a culture that supports excellence in all things, a strong and consistent message is required. This must

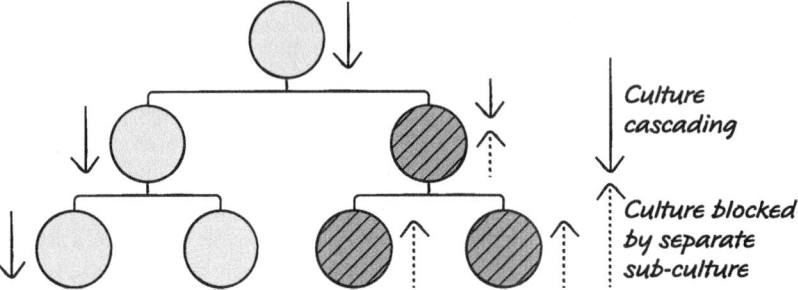

Figure 19.2 Culture blocked.

be delivered by a fully aligned team of leaders that has a good understanding of the desired approach; for whom it is a natural and comfortable way to operate and who are prepared to maintain their standards in relation to the desired culture.

In developing this cultural model and aligning the leadership it is very important to recognise that there is a lot of room for variety in a strongly aligned team. We are not looking for dozens of clones of the CEO who will all approach their jobs in exactly the same way. The preferred cultural and leadership traits are a very broad church and inclusive of many styles and approaches. They are high-level attributes that can be realised in many ways. For example, a particular company may have a set of cultural values that espouse trust, openness, excellence and learning. A given problem could be solved in such an organisation in a way which is highly creative or highly detailed and structured; by an individual or by a team; by trial and error or by getting it right first time. All while staying true to the values.

In no way does a consistent and strong culture prevent diversity in teams or approaches.

However, it is one thing to say that culture is cascaded down from the top of the organisation and another thing entirely to change it successfully. The process is simple, but it is not easy, as the existing culture that is in place will be deeply embedded.

Bottom up

Although there is a significant emphasis in business literature about leading culture change from the top, it is not simply a top-down process. Culture is an enormously complex area and notoriously difficult to get right. Like so much of safety, this is, in part, due to the myriad combinations of viewpoints from different people with different perspectives. A corporate culture is made up of many sub-cultures comprising the aggregated and intersecting beliefs, values and behaviours of a wide range of contributors. The idea that culture is entirely dependent on the management team is not only wrong, but smacks of the management 'we know best' arrogance that is so disliked by the work-force. Presumably, your company has had some success in the past, so the existing culture cannot be all bad.

So why fight it? Far better to embrace and incorporate what is already there. As the Romans moved their way across Europe over 2,000 years ago, they adopted and adapted local cultures. They built new temples over exist-ing sacred sites and incorporated local ritual practices into their religion, developing a fusion that worked for both parties. Some time later, Christian-ity used the same principles, usurping important local religious dates, which is why there are still pagan components in Christian festivals such as Christmas trees and Easter eggs. This process is nothing new. It is okay to want change to drive improvement, but in order to be successful this must be a joint journey, not a unilaterally imposed new state.

So, we must take the whole team on the change voyage. Change is also a complex subject and there are many books, consultancies and academic papers devoted to it. It is essentially an industry all of its own. Noting the irony of over-simplifying, it is possible to summarise that entire industry in three sentences:

1   By and large, people will resist change that is imposed on them.
2   They will cautiously support change if it is clearly explained to them in advance why it is happening and why it is necessary.
3   They will embrace change that they are involved in developing and driving.

For the most effective cultural change, therefore, the personnel already in the business need to be involved in the development of the new approach. The way to approach this is relatively straightforward, but remarkably uncommon in business.

Talk to them.

Sounds radical, but it works.

Go around the organisation, holding workshops, brainstorming sessions, surveys, online message boards or whatever is best for your business and ask them two questions. The first question is designed to find out what the key drivers for the workers are. What do they see as the *why* for the business? This is some variation on 'why are we here?' or 'why do we do this?' or 'what is it that makes you go the extra mile for this company?'

If you are in a good position culturally, the answers to this question arising from different areas will have similarities and may align themselves well with your stated company vision. It will then be easy to fit this into your new operating model. If you are not in such good shape, you may find a range of answers that are out of sync with the official line. If this happens, what your people are telling you is more likely to reflect the *real* culture. What is written on the company material is simply lip service. This demonstrates that the leaders in the business are not modelling their stated values, but some unwritten ones that their teams pick up on. Or that they are simply discon nected from the business.

The purpose of this question is to calibrate your leadership *why* with that of the rest of the business. It is well known and frequently demonstrated that teams perform best when they are in pursuit of a clearly defined common goal. This can often be seen during times of crisis. Everyone pulls together and delivers because it is really easy to see the goal. What are we trying to achieve? What are we all working towards? It is very easy for this to get lost in the day-to-day banality of our working lives. But if it is identified, captured and kept as a constant reminder, in a form that all of the teams have helped to develop, it becomes a powerful driver for performance.

The second question to ask is about behaviours. What are the behaviours you expect to see from a high-performing team? To help with this it is useful to imagine the colleagues that people would think of as high performers and determine what positive behaviours they consistently display. Or perhaps to look back at a project or task that went particularly well and identify the behaviours that underpinned that success. Not only do these examples make the concept a little more tangible, but they also keep those positives firmly rooted within the business. This helps to recognise the good work we already do and shows that the proposed changes are not too big a step from the current position.

The vast majority of people come to work and want to do a good job. They want to make a positive difference. This means that in any organisation, even the most currently dysfunctional, there are very positive behavioural traits that can be harnessed via this process. When people see these being valued and feel they are being listened to, they begin to engage strongly in the process. The key here is to take the positives that already exist and graft onto them those additional behaviours that are desired.

This second question will turn into a list of behaviours. If kept live, this becomes instrumental in embedding the desired approach. A clear list of behaviours, in the right environment, becomes a great tool to hold each other accountable. It provides a framework to help ourselves to perform and to challenge each other when we appear to be straying. In this way, the stated cultural values and behaviours become very real and are lived every day. This list will contain the good behaviours that your team have identified.

Hopefully, this will be a complete list of what you wanted to see. Almost inevitably, the list will include some versions of the following core list, simply because these are fundamental to a good workplace in every industry:

- transparency
- honesty
- good communication
- trust
- teamwork
- accountability
- professionalism.

Others may include less general ones that are more related to the specific business area – possibly creativity, passion, customer care or innovation. Pernod Ricard, the wines and spirits company, quote conviviality as one of their values that may inspire an entirely different set of behaviours. It is vital that these properly reflect your business. There is a lot of literature currently relating to innovative approaches to business, particularly in the hi-tech sector. Some of this is very interesting, but much of it is unproven as it is so new. New approaches are challenging convention, such as the opportunity to abandon a project for one that looks more interesting. This is founded on the basis that what is interesting to the team will be interesting to the market-place. Such innovation is a good thing, but while such a practice may work in creative industries, it's probably not going to be a suitable approach if you are designing aircraft. Not working on the landing gear because the auto-pilot looked more fun isn't really the rigour we would be looking for.

To get the most out of your drive for excellence, there are some other behaviours you will want to ensure are included. If they are not in your list, add them in. Do this openly. We're not trying to quietly slide them in and hope people assume one of their colleagues suggested it. It is perfectly acceptable for some of the behaviours to come from leadership as well as others from the workforce – it is a joint effort, after all. The following are essential for this to work (you don't have to use these exact words, of course, but don't stray too far from the meaning):

- a learning organisation;
- be prepared to challenge and be open to challenge;
- value and listen to the input of all staff;
- set and meet high personal standards.

It is critical that all people have the opportunity to provide input, know that they will be listened to and that we learn from that input and from the mistakes we make, as well as the successes we have. While we are developing as a business, that development is founded on every individual doing a great job, and setting and maintaining standards is the bedrock of excellence. Lots of small things being done well on a consistent basis, with a drive to keep getting better.

This development of an approach with the workforce needs to be genuinely co-created. We are not looking for the 'consultation' that usually accompanies changes where we are giving people the opportunity to comment but in reality all of the decisions have already been made.

Note that so far in our bespoke approach, we have not yet even mentioned safety. We are still putting in place the foundations that will support the disciplines required for good all-round performance. In this stage it is important that this process is rolled out for what it is – a way to improve our performance. By all means include safety as one of the areas that we want to improve, but do not push this as a safety initiative. There are too many ways in which people will see that as something being imposed on them and resist it. People generally think that they do the right things to be safe – whether or not those are in line with current company policy – because everyone has a different perception of risk and how it should be managed. However, pretty much everybody wants to be better *in general terms* and wants to see bottom-line improvements. Even if they think they're doing reasonably well, there is a 'good to great' concept to commit to. Quality of work, efficiency of production and profit are typically seen as having no upper boundary. Safety, however, is often viewed as having diminishing returns above a certain point, especially if you are in the fortunate (or possibly well-managed) position of not having had a significant injury for some time.

We're not yet looking for safety improvements. We're seeking excellence in the way we go about our business.

Keep the behaviours live by publicising them, talking about them regularly and by leaders modelling them consistently. Recognise and acknowledge good examples of positive behaviours. This will feel awkward and forced at first. But over time it will become a natural part of the business.

Review the output of this process against the initial thoughts of the leadership team and combine the two. Hopefully these will not be too far apart. Now that you have a clear sense of the values and behaviours to move forward with, you can begin developing those in a safety context.

Key points

- Culture is strongly influenced by leadership, but it is not only a top-down process.
- Develop a set of values and behaviours that work for your business.
- Genuinely engage with the workforce to co-create a culture that combines the best of the existing business with your aspirational components.
- Embed and live the values.

20 The system alignment

At some point, the business has to move from the leadership space into actually managing the work safely. This is the point when we become transactional and start actually doing stuff (including safety stuff). It is very fashionable to talk about leadership versus management and, given the amount of material singing the praises of leadership, one could be forgiven for coming to the conclusion that managers are bad for business. But management is hugely important. Without some structure and some organisation, businesses don't function and can't implement the vision of the leaders.

So, we need some structure. We need a system. And there is some irony given that in Chapter 14 we talked at length about reducing the amount of systems. But that was reducing them, not removing them.

The purpose of the system is to provide a body of knowledge so that people know what is expected of them and to provide a framework of repeatability and consistency across the organisation.

Unfortunately, if you wanted to be told what to put in your system, that isn't going to happen. Although there are some components that are likely to be common to most safety management systems (for which see the various safety management standards), it simply is not possible to develop an effective system without understanding the business and how it operates. Beware the person offering to sell you a ready-made safety management system. I'm working under the assumption that having read this far, you are looking for something worthwhile. I have seen too many examples of companies with systems and manuals that have clearly been written by an outsider with no thought to making it business-specific.

Such a system will never be properly implemented, will never get genuine buy-in and will never improve safety performance.

An external expert can be a big help in developing the system, but it should not be done by them in isolation. Ownership is vital. But it does mean that there isn't quite so much to include in this chapter.

So, rather than suggest how to go about developing or improving the system, here are some processes to go through; things to think about and hints, tips and pitfalls to bear in mind while doing it.

It should be noted that the development of the safety management system is one of the principal areas where businesses seem to put an awful lot of effort (excepting the copy-and-paste brigade). Many management systems are, therefore, well developed and comprehensive. A number are based on exactly the approach that will be described here. But in almost every instance, the management system is still seen as the controlling process for safety, rather than as an enabling process. If you already have a robust safety management system that aligns with this, focus on the implementation aspects to make sure that it is helping rather than limiting.

Structure and development

As we have discussed numerous times, the various different functions within the business need to interface and to be aligned in a holistic manner. There are a number of areas within a safety management system that also apply to other parts of the business. Investigations, for example, are frequently employed in understanding quality issues, as well as safety. Document management processes are important across all functions.

The safety management system should align with and interface with other functional management systems as part of an overall business management system. At some point, however, it has to be recognised that most of your staff have a functional specialism and work within a largely function-specific department. Most of them do not see across the whole business, so for your system to be effective it needs to provide both macro and micro information.

The system will be reviewed and amended over time to keep up with changes, but it should be largely consistent. For this reason, the structure should align to some fundamental aspect of how the business is operated.

Break the business down into a number of fundamental areas. This process will be very business- and industry-specific. You will be likely to have eight areas at the very least and want probably no more than about 16, although up to 20 is manageable. These should be the core aspects of the business. There is no right or wrong about how this is done, but it must reflect the business operation, whether by function, geography, product line or astrological star sign (not really).

Avoid breaking it down along organisational (i.e. line management) lines. Organisations can change, and do on a fairly regular basis. This structure needs to make sense irrespective of the current hierarchy and not have to be redeveloped each time you go through the inevitable centralise/de-centralise cycles (as a general rule, documents in a management system should have role titles rather than names when describing who does what, but as someone once pointed out to me, 'actually role titles here change much more frequently than people do' so it does need to be tailored to your own circumstances).

It should be based on your key business processes. What do you actually have to do to send widgets out of the door profitably?

For instance, an engineering and project management company may include:

- leadership
- design and engineering
- project management
- construction
- quality assurance
- safety
- risk management
- people
- commercial.

In that industry, these functions will always play an important role in delivering for customers.

Leadership is a must and most organisations will have safety, people, commercial and some sort of audit or review function. Other items vary. A manufacturing company may have 'operations' or, indeed, 'manufacturing' as one of theirs. Companies with mass-market products will have 'marketing' distinct from 'retail', while niche providers may have 'sales' which encompasses both. Public sector organisations may have a 'stakeholder relations' function. The point is that it summarises the key activities you undertake every day.

At the functional level, there needs to be a further split into different activity areas. Once more, these should cover the key activities that make up the overall function and this should come as no surprise. In safety, for example, this may look like:

- competence and training
- hazard management
- contractor management

- safety in design
- emergency response
- incident management and investigation
- environmental management
- rehabilitation (hopefully rarely needed)
- audit and review.

If your business has a strong process safety component, or large environmental risks, you may want to separate these out into their own areas.

Similar components can be developed in all of the functional areas by the relevant specialists within those areas.

These headings tend to look something similar to a traditional health and safety manual. However, the next phase will be different. A health and safety manual at this point usually goes on to describe what the business does in each of these areas. In this model, we are simply setting expectations at this point. For each of your identified components, develop a set of expectations that meets the company's aspirations for that area. In doing so, bear in mind all of the factors that we have discussed throughout this book – think critically about what expectation you are setting. Is it realistic? Is it motivational? How will it be received by the workforce? Are there any unintended consequences of its implementation? How does it interface with other functional areas?

Such expectations set an aspirational target that reflects the level you want to reach as a business. In setting them at this level, the vision for the business is given genuine depth and meaning. Most companies have visions that are relatively meaningless statements that sit uncomfortably on top of the rest of the business, balanced precariously on a shifting base of differing opinions, entrenched cultures and disconnected from the daily grind. By setting expectations in the way described, there is a clear line of sight from the overarching vision to the day-to-day preoccupations of most of the staff. A standard is thereby set that can be maintained through organisational changes, through peaks and troughs of workload and through the ebb and flow of safety performance.

Expectations may include the following types of statements:

- All accidents and near misses are investigated with a rigour proportionate to their potential.
- No staff may work on a task without supervision until they have been adequately trained and are competent to carry out that task safely.

These are pitched at a level that sets the standard, but allows the business the flexibility to deal with it appropriately.

In the first example, there is an allowance for small and low-potential events to be very quickly investigated, without wasting time and effort. But there remains an expectation that an event with high potential is investigated rigorously. This keeps the resources focused on the right place, but keeps an eye on all events. Contrast this with a statement that simply states that all accidents are investigated and defines a methodology. The business can change its investigation methodologies or tools without impacting on the higher-level expectation. A framework is set within which the business is empowered to operate.

In the second example, note that the requirement for training is task-specific. There is not necessarily a need to formally record proficiency before being allowed to do anything. Some tasks will be very low risk and so should be fine to carry on with. Others will require formal sign-off. Again, the business can decide how and why based on risk.

Involvement

The development of the system should be a joint effort with representatives from across the business. You can't involve everybody at every stage, but it is important that everyone knows there has been broad engagement and that as many views are considered as can be practically done.

There are two principal reasons for this, which are the same ones that are reflected in all the involvement and engagement activities we've already discussed. First, involvement fosters commitment. The success of the system in application is entirely dependent on the commitment to its use. Involvement engenders such commitment and results in people who are advocates of the system spread among the team continuously helping to maintain it. Second, involvement brings with it all the real-world nuts-and-bolts understanding of the business. Work as done, rather than work as imagined. There is little worse for the worker than seeing centrally imposed processes that don't reflect the reality of the work front.

Don't be afraid to ask for feedback and comment, or to run trials before full implementation. It is important to acknowledge that management does not know everything and is prepared to listen and incorporate suggestions. You don't have to accept everything, so long as the rationale is explained.

Scope

Think of the system as a framework to work within, rather than a set of procedures to follow. The high-level expectations that remain unchanging over time lead to freedom within boundaries, allowing the business to manage its own operations. These boundaries should be set where unacceptably high risk lies beyond.

This type of approach helps the business to think and speak in risk terms, which supports an enquiring and critical approach – not a mindless compliance with rules. It shows trust in the core skills and competence of the team to deliver on what they are good at – which is why you hired them, after all!

There is an element of compliance required, of course – hence the boundaries. There are legal requirements for safety that must be followed. There are also rules that are of benefit in those high-risk or highly complex areas we have already discussed. At the high level, though, the expectation should simply be that compliance is expected for all legal and company rules. The trick is then to clearly articulate those rules within working documentation only when there is a genuine need to make them. Overdoing it with the rules dilutes the importance of those that truly matter.

Intent

In many organisations, particularly those that are bureaucracy-heavy such as national and local government, over time the system itself appears to become more important than that which it was originally intended to achieve. A common example is a procurement system that requires multiple quotes for a purchase. This is intended to get better value for money, but in many instances there are insufficient quality suppliers of a particular service. A quote is sought and followed up to fulfil the quote requirement in full knowledge that the supplier will never be used, wasting everybody's time. Or for a small purchase the potential saving is grossly outweighed by the cost of going through the tender process.

I have seen plant management taken to task for not following procurement processes when purchasing a breakdown replacement part to get the process back up and running in the middle of the night shift at a weekend. The system mattered more than the outcome. Strangely, the procurement team agreed to a waiver process when they were asked for a call-out list so they could be dragged out of bed to be consulted the next time it happened.

By setting good expectations, the intended outcome is clearly articulated. This should be cascaded through all procedural documents by including a statement of intent in each procedure. If and when operational circumstances show that the detailed steps within the procedure are compromising the intent, welcome the challenge and review and update where required.

Presentation

A brief word about presentation. Use flowcharts, diagrams and visuals as much as possible. Many people struggle with long-winded wordy instructions, especially when dealing with field workers who are often not confident or welcoming of paperwork in any form. Although do beware of the great-sounding idea of a handy pocket reference. Many a pocket has been overwhelmed by the myriad of easy-to-reference cards from training sessions.

Review

Refer again to Chapter 14. Review your system on a regular basis and make sure it is still effective and appropriate.

Reviews tend to find problems and then add on solutions so that the system grows. Treat the review more like a pruning exercise. We prune trees to cut back unnecessary or dying branches and it allows the tree to flourish. Nothing ever flourished by having extra weight added on to rotten parts.

When you undertake reviews, don't fall into the old compliance audit trap that we discussed in Chapter 15. Your review should be a positive approach to identify improvement areas. The question to ask is not, 'Do we do what we said in the system?', but rather, 'Does the approach in the system meet the expectation or objective? If not, how do we deal with it in practice? How do we make it better to meet the objective?' All of these questions provide a much more open environment for improvement and learning. Traditional audits and reviews are poor learning tools.

Overall, your management system should be an enabler. It should help people to do their jobs, not hamper them by imposing unnecessary processes or insisting on procedures that don't achieve their intended objective.

Use it!

Of course, a safety management system is not simply a set of documents. There are far too many examples of manuals written as a management system and then parked on the shelf, never to be opened until the next audit or the next time to update it. For any system to work it needs to be implemented. If this doesn't happen, you don't have a safety management system, you have a work of fiction. And what is more, it's not even likely to be a good read.

A genuinely useful management system consists of not only the procedures, but also the people, the process, the equipment, the environment and the materials used. It includes leadership, training, communication, feedback loops, reviews, audits, inspections, observations and improvements. It requires understanding, a helpful user interface, visibility and profile. It must be risk-aware, smart but practical, thorough but flexible.

Which all sounds very hard.

But as we have discussed previously, safety is an emergent property of a complex system. It is not something you can easily target and improve. At the edges you can – when performance is very poor, it's easy to make improvements. There are many examples of companies achieving huge reductions in injury rates when starting from a poor baseline. But it is at the core where it becomes hard. The low-hanging fruit has been picked and if you want to see improvements you have to embrace the complexity. It is in these last few percentage points of improvement where it is difficult; where counting injuries is no longer a good approach to recognising improvement and where people get seriously injured or killed even though we thought we were doing the right things.

To avoid drowning in this sea of complexity, it is helpful to stop thinking about your management system in terms of a documented process. Maybe we should even stop calling it a management system – the very name implies that we can manage it, that it is knowable and controllable. But safety is a wicked problem. It possibly (almost definitely) has no single answer.

Instead of a safety management system, think of it as a *safety ecosystem*.

Safety ecosystem

An ecosystem is incredibly complex. It is made up of hundreds, thousands, millions of components that all interact and impact upon each other in innumerable small ways, aggregating to produce a vibrant whole. Yet within all this complexity, we understand our place. It is easy for us to comprehend and acknowledge the complexity even if we can't fully describe it. We can't really influence or direct it completely, but if we understand the principal core components we can at least stop misdirecting it quite so spectacularly.

A traditional management system attempts to codify all these connections, but it is almost impossible to get it all on paper. We do our best, because it is necessary to capture our knowledge and understanding somewhere, but it fails when we try to implement it, because it has constrained our thinking into straight lines, rather than networks. When we implement, think in network and ecosystem terms and develop an environment in which individual components can thrive by doing what comes naturally to them.

In an ecosystem, the birds, animals and plants don't have manuals to follow. They live by instinct, acting naturally in their small part of the ecosystem. Lots of small, simple processes build up into a cohesive whole that is often out of all proportion to the inputs in terms of its complexity and function. Beehives and ant colonies act as if there were a controlling mind optimising their output, simply by doing millions of small things in a simple way. This combination of many simple activities into a complex whole is a foundational concept in both evolution in biology and chaos theory in physics.

There are clear parallels with organisations. If we can create a safety ecosystem in which the safe option is also the natural and obvious thing to do, our output should improve. To do this we need to understand how people are hard-wired to think and make decisions and allow that to happen, rather than attempt to force them into thinking the way they are told to. We still need to recognise that unexpected outputs can arise due to the complexity, but we can begin the journey.

Thinking in terms of an ecosystem within which we operate, rather than a system which we manage, naturally brings considerations of individual psychology, teams, culture and leadership together with the more readily defined procedures and activities to think properly about how they all interrelate.

Consider the analogies in Table 20.1.

The safety ecosystem concept is not intended to be a tool to apply in developing your approach. It is a metaphor to help understand the interrelationships and the complexity. The fundamental point is to recognise the natural approach of people and incorporate it, which moves us into the realm of humans.

Table 20.1 Ecosystem–safety analogies

In an ecosystem	In safety
Individual elements require nutrition to thrive, but this comes in various types and delivery modes to best suit that element.	Each individual requires training and knowledge to carry out their role. We need to make sure this is tailored to both the individual's learning approach and the specifics of the role and its risk. Not one-size-fits-all training.
Elements carry out their role naturally, by instinct without having to deviate.	We need to develop work practices that are easy to carry out and take account of people's natural tendencies in thinking and behaviour.
Individual components can go wrong without overall damage to the ecosystem.	We need to design systems that are tolerant of individual errors.
Evolution encourages and embeds those innovations that are successful and discards those that are damaging.	We need to better understand variability in the work carried out and capture what is innovative and remove what is potentially dangerous.

Key points

- Build a safety management system that is designed specifically for your business needs.
- Make the management system a framework within which the business operates, rather than a manual outlining what to do.
- Engage the whole business in its development.
- Establish work processes where the natural and obvious thing to do is the safe way to do it.
- Be clear about the intent of the processes to prevent drift over time into bureaucracy.

21 The people postulation

We now have leaders on board who understand their role in supporting safety. We have fully engaged the workforce at all levels and we have a supporting system in place that enables work to be done, rather than restricting it.

The final piece in the puzzle, and possibly the most important, is to change our mind-set about the way we treat the people working in our organisations. The engagement process used in developing culture, behaviours and systems has already begun this change process.

More excellent work by Dekker (2014) looks at the way in which we deal with people in our investigations. In this he proposes an 'old view' and a 'new view' of human components in organisational systems. This has been copied and shared widely and further expanded upon by others (and by Dekker himself since earlier editions), but is summarised in Table 21.1.

In the overwhelming majority of incidents and accidents that I have seen or investigated, the individuals involved were trying to do the right thing to get the job done. On a construction site a fight occurred when the site safety adviser stopped a construction supervisor from unloading a delivery truck using a truck-mounted crane. The argument escalated. This had the two perspectives outlined in Table 21.2.

In both cases, the person was trying to do the right thing and from their perspective they were approaching the problem in the best way. Now, there

Table 21.1 Old view and new view of human systems

Old view	New view
Human error is the cause of accidents.	Human error is symptomatic of some other, deeper problem.
Complex systems are basically safe with humans as unreliable components.	Complex systems are inherently fragile and humans manage to operate safely despite that fragility.
Look for errors and judge the people who made them.	Determine the local context at the time that made the actions understandable.
People are the problem.	People are the solution.

Table 21.2 Differing situational perspectives

Person 1	Person 2
The delivery truck had been delayed and it was late afternoon on a Friday, but the driver had a long trip home.	Person 1 had no licence to operate a truck-mounted crane.
Person 1 had plenty of experience using a truck-mounted crane, although no formal training.	No-one else still on site this late had such a licence.
Using the crane to remove the equipment would get the job done and allow the driver to get away – reducing cost, effort and also making the trip home safer for the driver before he got too fatigued.	Rules state that uncertified people cannot operate machinery due to the safety risk of doing so.
	Production should not take precedence over safety, so even if there was a delay, that was better than having an unlicensed person operating the crane.

is no way that this should have led to a fight and there were some other situational factors that contributed to that. But the point is that there were many factors involved – there were time pressures and production pressures; there were different perspectives of risks; there were rules in place for well-intentioned reasons, although they may have been over the top for this instance (level ground, comfortable capacity, robust equipment, no people and no items nearby to damage with a dropped load); there were competing safety factors – risk of a dropped load compared with the fatigue risk of the delivery driver staying later – and so on. All of which combined to make a difficult decision for which two different people had different views as to the solution.

In this instance, it did not lead directly to an accident, although an altercation did occur. Every day, we place our workers in dozens of similar situations – there are conflicting opinions; competing priorities and pressures; complex systems and decisions; sudden changes in environments or work requirements; inadequate or unsuitable tools; insufficient experience; novel approaches or unusual circumstances that our procedures don't cater for and unpredictable people making unreasonable demands.

The real miracle is that, despite all of these issues and problems, we get through the vast majority of days without accidents.

People come to work intending to do a good job and the vast majority of the time they achieve this despite all of the complications that occur. The fact that they do is down to the remarkable capacity of people to be flexible, innovative and motivated to do the best they can to get the job done. And they generally manage that successfully.

There is a thread that runs throughout this book that forms the basis of the mind-set change that is required to improve performance. It is not a safety-specific aspect, but relates to all aspects of running successful businesses.

Fundamentally, this is to stop thinking of people as a problem to be fixed (old view) and start thinking of them as a resource to be harnessed (new view).

We hire people because we think they can do a good job, and then we treat them as if they are stupid and untrustworthy. Find out about your people and what skills they have to harness. Across industry we have people working in 'unskilled' roles who are governors at their children's school; we have capable, highly qualified and experienced people having a break to raise their children and then returning to work in a role several levels below their capacity; we have staff with no authority to buy station-ery who are treasurers of the local operatic society and we have another having to call an IT helpdesk to fix the most basic issue on their computer even though they run a web design business in their spare time. The list goes on.

Use these people and harness their capabilities to improve. We can do that by understanding how their normal day works and where difficulties, frustra-tions and good practices lie.

Learning from normal operations

In the late 1990s a scandal emerged involving the manufacture of nuclear fuel by British Nuclear Fuels, the UK government-owned nuclear services company. Ceramic fuel pellets made for a Japanese customer were supplied to a specification that included physical dimensions for each pellet. As part of the manufacturing process, an operator would measure the pellet and record the size. Subsequent analysis of a batch of measurements provided showed that the sizes did not have a statistical distribution that would be expected from the production process used.

The investigation into this apparent data anomaly showed that measure-ments were not being taken and the data was falsified. The measuring of pellets over and over again was time-consuming and boring. Measured pellets were never found to be outside of the specification, so the operators recorded fake numbers to save themselves the bother of having to do the work.

It is human nature to look for a shortcut when a task is tedious. If the task also seems to serve no apparent purpose, because the measurements were always within specification, there is even more incentive to cut corners. Whatever the right or wrong of the operators' actions, the final outcome was the reworking of all the suspect fuel batches, at great cost, and a massive rep-utational blow to the business. Although this was an incident relating to quality assurance rather than safety, there are definite parallels.

In traditional safety terms, we wait for an accident and then we investigate. We find out what we think the operator did wrong, probably fire them, and then warn everyone else against doing the same thing. We (possibly) prevent it from happening again but we have had to endure the consequence of it going wrong in the first place. To alleviate this, most organisations will record and investigate near misses. Even though we don't actually have an injury before investigating, we have still lost control and it is mostly luck that has prevented the injury.

What if we went a step further and investigated when seemingly nothing had gone wrong? What if we could learn from normal operations and antici-pate and prevent problems well in advance of them becoming an accident?

The vast majority of the time, workers get things right. As we have already discussed, they have to deal with hundreds of different factors in their daily work and they do this successfully. In fact, if they did not do so, the job would never be done. Humans are the adaptive part of the system that enables it to work. Most of these daily variations are fairly mundane. A minor tweak here or a change in the order of activities there helps achieve the final outcome. But occasionally, they are innovative and clever and occasionally they are unsafe and increase risk. If we can learn from normal operations, we can capture those that are innovative and share them, while finding those that are riskier and redesigning the task to take away the motive for doing them.

Every operation will have its pinch points, its pressures and its difficulties that carry within them the seed of a future accident – learning from normal operations allows us to remove them long before they eventuate. It has the added advantage of increasing efficiency as the majority of items identified tend to be barriers to effective work as well as safe work. It also enables the capture and normalisation of innovation.

When we look at normal work, we need to ask the workers in order to get their perspective on how work is done compared to how we imagined it when we wrote the procedure. This can only be done in an environment of trust where they feel comfortable revealing their shortcuts. The process is very simple. Just be curious and ask questions, such as:

- When is work difficult to complete?
- What frustrates you while doing the work?
- What does a good day look like?
- What does a bad day look like?
- If there was one thing you could change about this activity, what would it be?

This brings out the information needed to either harness innovation, or to identify risks that the system has introduced. But it also acknowledges the expertise and experience of the workers and gives them some control over their working environment; both of these are highly motivational and bring broader benefits than simply fixing the immediate issue.

Having identified issues in normal work, the workers are also well placed to develop solutions. They have both intimate knowledge of the process and the ability to immediately recognise the limitations of potential changes. Health and Safety Committees can be re-imagined as work design teams. Lessons learned reviews can evolve from the currently prevailing 'what went wrong' reviews to genuine learning and improvement-focused reviews. As well as pre-start meetings, we can introduce post-completion meetings that look for these learning events.

Had management asked similar questions about the pellet measurements, perhaps they would have foreseen the potential for the problem and implemented an automated process or some alternative method, avoiding the cost and reputational impact of the falsification.

Available information

Workplaces have become safer over time. The best safety performers now have very few accidents. We talk about accident rates in bland numerical terms, but actually workplaces are, on the whole, incredibly safe places for the vast majority of the time. Roughly speaking, people work around 2,000 hours per year, over a 50-year career. This is 100,000 hours per career. An accident rate of one per million hours therefore equates to approximately one accident for every ten careers. This is pretty safe. I think most people would agree that working for ten lifetimes before an accident occurring is acceptable.

What this means is that, by restricting ourselves only to accident data for learning, we have very, very little information available to use. We can increase this by an order of magnitude by investigating near misses as well, but it remains small. In any case, the events that cause fatalities often don't have any obvious near-miss precursors. Reported near misses will often not include those that involve shortcuts or workarounds as workers are less likely to report something where they were not following a procedure.

Learning from normal operations, however, opens up every hour of every day for useful information. There is so much more to learn, it just takes a little more effort and engagement. In fact, the challenge becomes one of practically dealing with such a huge amount of information rather than getting enough to be useful. Here, technology can help with tools now available for collating real-time feedback from large numbers of people. Imagine a web meeting where large numbers of staff can text in their single biggest frustration to give an immediate top-ten list of improvement activities.

Being statistically unlikely is little comfort to you or your family if you are the individual who is seriously hurt or killed. So, we owe it to ourselves to use the information that is available through normal operations to learn and improve.

Failure tolerance

Having rightly praised the workers for their capacity and capability, it should also be recognised that they will make mistakes. Failure of the system is inevitable. This may be related to the workers' roles or it may be related to equipment, process or some other factor. But reality simply will not let it be otherwise.

Humans will make mistakes, equipment will fail and circumstances will combine to create accidents. In many instances, we can predict and prevent the failures, but at some point an accident *will* occur.

In the development of systems and approaches, it is vital to ensure that we build in the capacity to cope with failure when it occurs. When the nuclear reactor at Three Mile Island failed in 1979, no radioactivity was released to the environment because the event was a predicted possibility that the containment systems were designed to accommodate. In the significant coverage of the event that followed, the engineering capacity of the containment was largely overlooked, but from a safety systems perspective that part, at least, was a complete success.

In high-risk industries such as the nuclear, aviation and petrochemical industries, there is a highly evolved and rigorous safety engineering approach that deals with equipment failure. Components are analysed and their likely failure modes determined. Redundant equipment is added as a standby and indicators and monitors are installed to identify operating conditions that may point to potential failure modes. Failure is expected and catered for. Compare this to the majority of activities that are largely worker dependent. We prepare an instruction and simply expect the workforce to follow it – anticipating correct actions every time.

When building our systems, we should do so to cater for when an accident will happen. Earlier on, we discussed the need to clearly identify and focus on high-risk activities. This focus should include the question of when this system inevitably fails, how well placed are we to cope with that failure?

Key points

- Treat people as a valuable resource and source of solutions, rather than a source of problems.
- Recognise that your system is not fail-safe and design it to be tolerant of failure when it does occur.
- Learn from normal operations to identify variability in processes that either needs to be dampened or encouraged.

Reference

Dekker, Sidney (2014) *The Field Guide to Understanding Human Error*, 3rd edition, CRC Press.

22 The future evolution

It has been pointed out a few times that there is good knowledge and information available out there that appears not to be making it to the majority of operational safety professionals.

This chapter provides pointers to a few of those sources. There are many others out there, some of whom are renowned academics, others simply smart and thoughtful practitioners. There are a number of online forums where interesting discussions take place. In some cases across this knowledge base, there is a degree of intellectual navel-gazing and it can, at times, be difficult to see how the interesting theoretical ideas can be translated into practical, working solutions once we take into account time, budget and local knowledge. We could, for example, develop new approaches to investigations that involve profound understanding of human behaviours and interacting systems, based on the work of psychology professors combined with chaos theory analogies. This is all very well if you are investigating an exploding space shuttle and you have millions of dollars and a large team, but it is not particularly helpful if you are on a building site with safety as a part-time responsibility, investigating why a scaffold collapsed.

While acknowledging this difficulty, there are many benefits to research and theorising for their own sake. This is often how innovation occurs and knowledge advances in sometimes unexpected directions. The trick is not to criticise the academic and theoretical nature of this, but to work out ways that it can be realistically applied. The ability to translate between academia and practicality is a rare skill and its scarcity is one of the fundamental reasons why safety theory has not advanced more quickly into routine application. The more real-world practitioners are exposed to the theoretical concepts, the more likely we will find people capable of making that translation.

The pointers below are a combination of theories and specific people that may be of interest. Some of these have already been referenced in relation to specific publications.

Safety Differently

Safety Differently is less of a theory than it is a philosophy. A range of posts and articles can be found at www.safetydifferently.com. The site was established by

Daniel Hummerdal from Australia and is edited by Ron Gantt in the US. Ron's articles are particularly insightful, thoughtful and well-written. It draws strongly on the work of Sidney Dekker (see below).

The Safety Differently philosophy is espoused in its name. It is simply an effort to do things a little differently by challenging some established approaches with some fresh thinking and fresh perspectives. It has morphed somewhat into a 'movement' that runs the risk of people seeing it as an approach in its own right. This could not be further from the truth. Inherent in the philosophy is the idea that change and challenge are positive enablers in development. This means that the idea of a fixed way of doing things is neither desirable nor beneficial. There is not, therefore, a 'safety differently' way, except in the way that challenge occurs (note that this is my interpretation of the idea behind Safety Differently).

Safety I and Safety II

Safety I and Safety II is the terminology used by Professor Erik Hollnagel, a Danish academic. Fundamentally, Safety I incorporates 'traditional' safety approaches and Safety II is a more modern alternative. In Safety I the focus is largely on negative aspects – examining what went wrong in the event of an accident and then trying to fix it. Hollnagel's view of a more productive approach (Safety II) is to focus on the positives – understanding what went well when something was successful and implementing it more widely. Essentially changing from a reactive, accident-based approach to a proactive, defence-based approach.

There has been some discussion as to whether Safety II is a replacement for Safety I or if it is an addition or an evolution. This is probably over-simplifying the viewpoint. No doubt there are aspects of traditional safety that can still be valuable and, let's face it, we're unlikely to ever stop investigating accidents and trying to stop what went wrong. In any event, Hollnagel describes it as complementary.

Many organisations do have a lessons learned process that could loosely be described as reviewing what went well, but they are typically haphazard and badly implemented, with the positives usually something of a sideshow to the main event of finding what went wrong and improving it for next time. What Hollnagel is describing are much more thorough and formal investigations into positive outcomes.

Hollnagel is also responsible for the ETTO principle (efficiency–thoroughness trade-off), stating that we can usually either be thorough or efficient but not both, typically based simply on available time to complete a given task.

Systems thinking

Systems thinking examines the structure and performance of systems, where systems are defined as a collection of components that all interact. Each component adds to the system to make a complex and cohesive whole, but the output of the system is dependent on not only the individual components, but also the myriad of ways in which those components interact with each other, making it dynamic and difficult to predict.

This is essentially the basis of the ecosystem discussion previously. In safety, there are many interacting components that may include human aspects or more abstract aspects, including:

- management
- regulators
- workers
- unions
- cost pressure
- time pressure
- government
- society
- location
- environment.

The seminal work on systems thinking in safety is by a professor at MIT, Nancy Leveson (2012). This is a fairly major piece of work to go through, though, at over 500 pages.

The basic concept in systems thinking is quite obvious, but is something that we usually miss. Simply put, it is not possible to predict or explain behaviours and events fully without considering all aspects of the system in which they occurred. A perfect example of this in safety is the influence of the regulator over businesses. Over all aspects of safety management hangs the shadow of the threat of legal action. Whether it is a worker carrying out a task, or a manager defining what needs to be done, the impact of that threat always needs to be considered when explaining why a particular course of action is taken. We put far too much store in bureaucracy to prove we have done something due to the potential need to justify it in a future prosecution. This is due to the impact of the regulator, even when they are not present and in any event the likelihood of prosecution is pretty slim. This interaction, therefore, explains some behaviour that we take. In a similar way, if we want to effect change in that record-keeping mind-set (see Chapter 14) we cannot do it without considering (and probably changing) the role of the regulator. Although, in fairness to the regulator, safety professionals do have a tendency to implement what they think the regulator wants, rather than it being anything specifically stated as required.

Human factors

Human factors is included here less as a cutting-edge, academic discipline (because it has been around for decades) but more to demonstrate the difficulty we have in translating from concept to practical application.

Human factors as a discipline consists of a large body of work and concepts. It considers the many roles of the human in work systems and environments. It includes areas such as:

- incorporating ergonomics into engineering design to minimise the potential for accidents or for physical strain;
- recognising human behaviour in the design of systems to make them tolerant of how people think and act;
- optimising human–machine interfaces to maximise performance and minimise failure;
- understanding likelihood of different human error types;
- recognising, predicting and preventing error-producing conditions, such as fatigue, pressure, stress, cognitive overload, complexity and others;
- the effect of individual, group and organisational psychology on behaviours.

It is a fascinating area that provides many useful insights into how we can improve safety but, while the name is bandied around, its application in most organisations is limited to ergonomic assessments of workstations and possibly some basic behaviour-based safety tools. Yet these human factors considerations have been around for years. One of the very first conversations I had at work following graduation over 25 years ago was with one of our human factors specialists taking me through a human error decision tree process in order to feed some operator error data into a fault tree to calculate the frequency of an accident.

Even after decades of research and work understanding the reasons for people's actions, we still spend most of our time saying that a person behaved stupidly (with hindsight) and that most accidents are nothing that can't be sorted out with a healthy dose of common sense. This is symptomatic of safety's inability to practically apply good-quality thinking.

All safety practitioners should have a healthy knowledge of proper human factors considerations. At the very least it should be widely recognised that human factors cover not only the actions of humans in the system, but also the impact of the system on the humans within it.

Individuals

Sidney Dekker is a professor at Griffith University in Australia. He has written a number of books and papers, as well as being an accomplished and provocative speaker, so a search on YouTube can be a very fertile source of good thinking.

Professor Andrew Hopkins of the Australian National University has a sizeable body of work with a strong emphasis on major accidents and the organisational failures that contributed to them, particularly well known for reports following the Longford and Deepwater Horizon explosions.

James Reason is a professor at Manchester University in the UK. Best known for his 'Swiss cheese model' of accident causation, he has produced a much larger and very varied range of publications that are well worth exploring.

Todd Conklin has worked for over 25 years at the Los Alamos National Laboratory in the US. He presents a regular podcast providing insightful and challenging views on a wide range of topics.

Reference

Leveson, Nancy (2012) *Engineering a Safer World: Systems Thinking Applied to Safety*, MIT Press.

23 The highlights

We've all been away on a training course and left the venue brimming with enthusiasm and a myriad of ways to improve when we get back to work, only to find it is all but a fading memory after a few days back in the office faced with the tyranny of normality. It is unfortunately all too easy to fall into this trap. Reading things in a book can result in the same problem, although there is the advantage of being able to revisit it reasonably frequently.

There are a number of concepts and discussions in this book that may help improve safety performance. But only if they are applied. This may take some courage and some leadership, depending on your workplace, or it may be welcomed with open arms. Clearly, it is far easier to talk about these things than it is to do them.

In recognition of the difficulties in making sweeping and wholesale changes, the following list of highlights gives those areas which will make the biggest change – noting that some of these may still be difficult to do given the cultural and behavioural changes required. If you do nothing else when you put this book down, work towards these.

People

Recognise the amazing job that the workforce does every day in dealing with ambiguity, change, pressure, constraints and contradictions. Find out what they do when things go well and work out how to replicate it and build on it. Use their knowledge and trust their skills.

Find ways to bring safety back into normal operations and learn from those normal operations. Redesign your equipment and processes to make the natural and obvious action the safe one to take.

Your people are a resource to harness, not a problem to control.

Measure safety, not unsafety

Count and measure what we do that makes safety happen and stop relying on counting when we are unsafe. Especially lost-time injuries! Remember that lack of injuries does not equal safety.

Provide reporting processes that are informative and useful, that don't summarise so far that all meaning is lost.

Own safety

Safety is the job and the responsibility of the leaders of the operations carrying it out. Don't rely on the safety manager to do everything. The person who creates the risk owns the risk and must manage the risk.

The safety team should be coaching, advising and supporting, only taking the lead when there is a particularly specialist problem to be resolved.

Build expertise

Acknowledge the complexity of safety and build expertise in understanding it. All levels of management should have a good understanding of the deeper and more complex aspects of safety to allow them to properly manage it and to take the appropriate actions. Only by properly understanding the complexity can you find ways to make it simple in application.

Treat safety with the same level of rigour and critical thinking that you apply to the financial aspects of your business (see Appendix 4).

Focus on higher risks

We only have a finite amount of resource to work with. Identify, understand and manage your high risks first before diluting resource going after the lower ones. Where else in your business do you deal with the small stuff and hope that the big stuff will look after itself as a consequence? It is 'penny-wise and pound-foolish'.

For every action that you take and for every process that you put in place, ask yourself how much it is doing to directly reduce risk exposure.

Major accidents are rare. Unless we have a clear and consistent focus on high-risk events, they will take us by surprise and we will continue to repeat the cycle of major accident – reform – forget – major accident.

24 The Resistance

We have explored a number of myths and failings, as well as alternatives and solutions. You should now be armed with the ammunition to challenge and change for the better. Unfortunately, in this fight you will meet up with The Resistance. The rear-guard of the old guard.

Safety has undoubtedly improved significantly over the last century or so. The majority of obvious failings have been designed out, or safe systems developed to protect workers. Vehicles are much more robust than they used to be. Carefully planned task analyses are undertaken. Machinery is properly guarded. Fall protection is provided when working from height. Automatically initiated interlocks prevent process failures developing into major consequences. Although, sadly, there are still workplaces that have not yet come to terms with all of these improvements.

The pace of change has been rapid over much of that time, but it now seems to have ground to a halt. Partially, this is because the easy fixes have been done. As in many endeavours, the final few percentage points of improvement are the most complex and the hardest to make. Much of the remaining improvement requires changes to the way people behave, not in a blame-the-worker-unsafe-act way, but in a communication, interaction, management, leadership, motivation, cultural way. Very little is more complex and difficult to improve than people.

Yet despite these inherent difficulties, there is a lot of research available, a lot of studies undertaken and considerable amounts of experience in their application in the field. So what is happening? Why is this information not reaching workplaces? Why are safety professionals not having challenging conversations with their boards, executives and staff about cultural factors and motivation? Why are we still trotting out flawed ratios from decades ago – even teaching them to the next generation of our profession in a never-ending perpetuation of analysis-free thinking?

The information that is out there is only in safety research land. For it to become entrenched in the workplace it has to be understood and championed by boards and executives and cascaded into the workforce. The ubiquity of the zero paradox proves that this can be done, albeit erroneously in that case. Part of the reason for this 'success' is that it is a simple, catchy idea

that has superficial appeal. Unfortunately, in the modern world of short attention spans and competition for thinking time, superficial and appealing is
always at the forefront of adoption. Safety is not, however, simple, as we have
discussed at length.

We are guilty as a society of dumbing down to gain acceptance and safety
is not immune to this. Ironically, we make safety complex at the work front,
where it should be simple and simple in the boardroom, where it ought to be
given the thoughtful consideration it deserves.

Academics are well known for their inability to apply their knowledge to
'real-world' applications – sometimes justifiably, sometimes less so. The
majority of workers that are routinely exposed to high risks in their roles are
renowned for their disinterest in, and distrust of, paperwork. There is a small
group of people who are capable of dealing with both camps and can help
translate theory into practice. This group needs to be conversant with the
language and the culture of both sides of the divide. Consider engineering,
where a substantial proportion of the profession spend most of their time
dealing with the nuts and bolts (literally) of reality, yet have a strong academic
background. There is ample feedstock for the translators that can take cutting-
edge research and apply it practically.

In the safety profession, there is a substantial proportion with strong practical backgrounds, but with little or no academic bent. There are too few
bringing new theory into industry for it to reach a tipping point and develop
widespread acceptance. Some industries are better than others and these are
typically those with the potential for major consequence – aviation, nuclear
power and the like, but even there, outside of the specialist safety engineers
(and sometimes even including them), is a general lack of understanding of
some of the more subtle and complex aspects.

In the presence of this vacuum of knowledge, practitioners resort to using
simplistic tools – rules and compliance – while management embrace overly
simplistic slogans and sound bites.

Note that this is about the ability to think critically, assisted by some academic experience in a suitable field. It is not about insistence on a 'safety'
qualification – particularly as many of those currently available perpetuate
some of the poor thinking. Nor is it intended to imply that only degree-
qualified academics can think critically. This is clearly not the case, but the
rigour of a strong academic background does instil some of the disciplines and
challenging processes that support it. There is a need for the safety profession
to have both practical and academic types, but right now the balance is
wrong.

For these different approaches to become accepted into the mainstream
they have to travel from academia to industry. This transition is filtered via
the safety profession. But currently the filter is blocked with years of assumptions and habits with insufficient reflection to understand that this is a
problem. As well as a significant proportion of the profession who are not at
ease with, or aware of, academic research.

One of the biggest challenges of the safety professional is attempting to change the habits of a time-served workforce. 'I've been doing this for 30 years and it's never caused me any harm' can be a difficult argument to counter and every safety professional will tell you they argue against it all the time. Yet, those very people are using that same rationale to dismiss or ignore progression in their chosen field.

> I know that most men, including those at ease with problems of the greatest complexity, can seldom accept even the simplest and most obvious truth if it be such as would oblige them to admit the falsity of conclusions which they have delighted in explaining to colleagues, which they have proudly taught to others, and which they have woven, thread by thread, into the fabric of their lives.
>
> (Tolstoy)

It is very difficult to reflect on, and challenge, your own attributes, capabilities and approach.

So, there is a strong resistance to change within the industry. Not change per se, as the majority of safety professionals are continuously trying to effect change, but change related to the fundamental principles on which our current approach is based.

How do you recognise The Resistance? By far the simplest way is to analyse the arguments made in support of, or against, a particular approach. This is the same in any walk of life and argument. The critical thinker will ask probing questions, look for evidence, draw conclusions based on logic and reasoning. They will understand the role subjectivity and emotions play in dealing with people, but recognise that as one factor in the argument. The Resistance will often argue in purely emotional terms, or very subjective terms based on their own direct experience with little corroborating evidence from other people, other industries, or different approaches.

The critical thinker will consider interfacing areas; look at long-term as well as short-term implications; think about unintended consequences; be open to changing approaches if evidence shows a requirement to do so. The Resistance will focus on the specific issue at hand, be certain of their own conclusions and speak in absolutes.

The most obvious example is, as we discussed earlier, around zero harm. When challenged with the potential for unintended consequences, the most frequent response is: 'If it's not zero, how many people are you planning on hurting?' An absolute statement – no shades of grey allowed – with an emotional foundation.

When discussing safety as a priority, The Resistance speaks of safety 'having to be number one'. Another absolutist position that doesn't recognise that there are other requirements out there and that there are broader issues at play.

Whenever you listen to a debate on any subject – politics, religion, homeopathy, which team deserved to win the game – the same patterns will often emerge. One party will be reasoned and critical, the other relying on emotions because their argument lacks substance.

At times, there is little that can be done to sway people that have a fixed view. Consider two strongly religious people from different persuasions. No amount of argument, evidence, positioning or logic will move them from one side of the debate to another. The belief is so ingrained and so deep that they will develop all sorts of highly tortuous and tenuous justifications to twist the argument in their direction – even in the face of the most compelling logic.

But, as we have said, there is no definitive right or wrong in many instances of safety discussion. As such, it is also possible for critical thinkers to fall on both sides of a debate. In that instance it is occasionally refreshing to see complex and difficult problems deconstructed; to see people prepared to shift their positions in light of new facts and well-supported views and for both parties to learn something for their betterment. This is, unfortunately, a relatively rare occurrence.

People can, however, change incrementally over time when immersed in the right environment as cultural factors shape their thoughts. It is the responsibility of the senior leaders within industry to ensure that they, and the senior safety managers who support them, apply the appropriate critical thinking capability when developing strategies, messages and structures. This will set the tone and direction for others to follow and, over time, will lead to the improvements that are required. Those incapable of change, even via this slower and more gentle approach, must eventually fall by the wayside.

We started this discussion at Pike River Mine in the South Island of New Zealand. Let us only go back there to remember and reflect, not to repeat.

Appendices

Appendix 1: managers' checklist

As a manager you should be, and you probably are, committed to the safety of your people. In fact, I would go as far as to say you definitely are, otherwise you would not be wading through a book about it.

But how well do you demonstrate that commitment? How does it manifest itself day to day within your organisation?

Look through the questions below and ask yourself how well you're doing? Can you identify actual examples of when you demonstrated the particular behaviour or approach? And no matter how well you are doing, how could you do better in the future? Think about whether your workers would give the same answer as you do – maybe even ask them for feedback.

Once you have read through the book, you will be able to add a number of other questions to this list.

- How visible am I in the workplace (not necessarily just from a safety perspective, but generally)?
- How do people know about my commitment?
- Do I follow up after conversations when safety issues are raised?
- Are my words and actions on safety always aligned?
- Do I include safety as a matter of course in all work activities or only discuss it when it is an agenda item?
- What is my own knowledge of safety like? Am I fluent in the language and use of safety tools and processes?
- Do I know what the principal hazards are that the team face and how they should be managed?
- Do I ask about safety, rather than telling about it?
- Do I know what safety roles and responsibilities different people have?
- Do I get involved in incident investigations?
- Do I actively share lessons learned?
- Do I set, demonstrate and uphold consistent and high standards and follow through on commitments?
- Do I fix things if I can and offer my help to address other issues?

- Do I help translate concerns into actionable ideas?
- Do I ask for (and act on) ideas on how to improve?
- How do I recognise and reward good safety behaviours?
- Do I reward and recognise people for raising concerns?

Appendix 2: a day at the races

People look for patterns and explanations in numbers and often don't think later-ally or critically about them, their sources or their implications. In a world where we have to manage risk, people are particularly poor at understanding the role of chance. It is common to hear phrases such as, 'We're due an incident according to the law of averages.' I once heard a risk manager say that they didn't want to focus on events that had too low a likelihood (less than once in ten years) as, 'There is no point worrying about things that won't happen in the reasonably near future' as if the event happening was on a schedule rather than a probability.

People also take into account their own perceptions when interpreting information, removing objectivity. In his show, British illusionist Derren Brown set up a horse race betting scenario using the fact that people often believe there is a 'system' or 'inside track' in betting that can be exploited. Together with their assumptions about patterns and trends Brown was able to make someone part with their life savings.

If you have never seen Derren Brown, it is well worth looking up his show (and reading his books). Not only for the entertainment value in what he does, but also in demonstrating psychological frailty of people and reframing expectations about the sort of behaviour that can be expected from them. While obviously not directly dealing with workers, it does demonstrate how predictable people are in their behaviour – just not in the way that most people think. This means that so many of our assumptions about people and their behaviour are wrong. This is directly applicable to many of the concepts we are discussing here. Go to www.derrenbrown.co.uk for more.

On this particular show, the process went something like this:

A person was randomly selected to receive horse-racing tips based on Brown's system, which he claimed would predict winners. The horse won.

For the next race they received another tip. This horse also won.

The race after: another tip, another horse, another win.

As this progressed, the recipient became more and more convinced that there must be some truth in the predictions. The winning streak continued until one day she was taken to the races and told to get together as much money as possible to put on the next tip.

Convinced that the process was foolproof she proceeded to collect as much money as she could, including borrowing some from her father.

Based on the trend she had seen, she placed £4,000 – money that she didn't really have – on a horse. Seeing the pattern (and also knowing that Derren Brown was involved, which was revealed prior to the last race) was enough for her to risk everything she had at the races.

Questions she didn't ask:

- How else could this trend be caused?
- What does a historical trend tell us about the next event?
- What underlying information should I know in order to trust the trend?

Actually, she did ask herself some of these questions after a fashion, but could not really articulate clear answers.

So how did Brown do this? He simply took advantage of a knowledge of probability and applied it to people's perceptions. This is what happened (the numbers are made up for illustration).

- For the first race, an initial tip was sent out to 250 people. This was repeated *for every horse in the race*, giving a total of, say, 1,000 tips covering four horses.
- After that race, a second tip was only sent to that proportion of people who had won in the first one, 250 people. Fifty were sent a tip for each of five horses in the race.
- After this race 50 people got a third tip, again spread across all runners.
- After each race, most people fall away, having lost, until only a few are left and eventually someone has a five-race winning streak.

From the outside, it is obvious that there is no prediction involved, but from within, you only see your own story, not realising that you are just part of a large sample.

Of course, it is very easy to sit in your armchair (or desk chair depending on where you read your safety books) and say, 'That's all very interesting, but I would never fall for it.' The trouble is that social experiments continually show that susceptibility to this kind of thinking is far more prevalent than we generally think. So, even if you are numerate and protected, those around you may not be.

We do this all the time in safety – misinterpret data and act on headlines without detailed analysis. Contrast this to a situation where, perhaps, a single worker is involved in three consecutive incidents. In many situations, even with companies using a 'just culture' process, the worker will now be going through disciplinary action. But the trend shown could easily be due to other factors rather than some sort of assumed negligence on their part. Perhaps they report incidents others won't and they are in fact simply indicative of the reality of the workplace; perhaps they are normally the most reliable worker and so repeatedly chosen for the most complex and highest-risk task, or perhaps they are simply unlucky and, left alone, would spend the next ten years incident free.

Appendix 3: how bad policy happens

The following light-hearted criticism of corporate communication has been floating around offices for years. I have tried to trace the source, but have been unable to. If any reader knows where it comes from, I will be happy to attribute it.

In the beginning was the plan.
And then came the Assumptions.
And the Assumptions were without form.
And the Plan was without substance.
And darkness was upon the face of the Workers.
And they spoke among themselves, saying,
'It is a crock of shit, and it stinketh.'
And the Workers went unto their Supervisors and said,
'It is a pail of dung, and none may abide the odour thereof.'
And the Supervisors went unto their Managers, saying,
'It is a container of excrement, and it is very strong, such that none may abide by it.'
And the Managers went unto their Directors, saying,
'It is a vessel of fertiliser, and none may abide its strength.'
And the Directors spoke among themselves, saying one to another,
'It contains that which aids plant growth and it is very strong.'
And the Directors then went unto the Vice-Presidents, saying unto them,
'It promotes growth, and it is very powerful.'
And the Vice-Presidents went unto the President, saying unto him,
'This new plan will actively promote the growth and vigour of the company, with powerful effects.'
And the President Looked upon the Plan, and saw that it was good.
And the Plan became Policy.

Appendix 4: critical thinking questions

I use the term critical thinking fairly loosely in this book. But actually, it has been developed into a particular set of strategies and approaches through research and it has developed into an area of expertise in its own right. It is worth exploring to understand the underpinning principles and benefits from formal application of critical thinking techniques; however, for the purposes of our thinking here, the following list of questions provides a shortcut to some of the right levels of thinking required in its application to the safety context.

Some critical thinking questions:

• What are the implications in the short term, medium term and long term?

- What interactions occur with other parts of our system, equipment, people, business?
- What assumptions have been made in coming to this point?
- Are those assumptions valid?
- Are there any biases that may have been involved in coming to this point?
- How will this be interpreted by other stakeholders?
- Will this achieve what we want it to?
- Will there be new risks or issues introduced – unintended consequences?
- What if it fails?
- How tolerant of error is it?

Index

Page numbers in *italics* denote tables, those in **bold** denote figures.